The Investigation and Management of Erosion, Deposition and Flooding in Great Britain

London: HMSO

ISBN 0 11 753117 0

Acknowledgements

This report was prepared as part of a research project entitled **Review of Erosion, Deposition and Flooding in Great Britain**. The project was funded by the Department of the Environment under its planning research programme (Contract No. PECD 7/1/412).

This report was written and compiled by Mr. E.M. Lee of Rendel Geotechnics, with guidance provided by Dr. A. R. Clark (Rendel Geotechnics) and Mr. A. S. Freeman (Rendel Planning and Environment). Specialist contributions were provided by:

●   Prof. M. Newson (University of Newcastle)
●   Dr. J.C. Doornkamp (University of Nottingham)

Further specialist advice has been provided by:

●   Prof. D. Brunsden (Kings College, University of London)
●   Prof. J. Pethick (Institute of Estuarine and Coastal Studies, Hull University).

This review has been undertaken in order to provide **an overview** of erosion, deposition and flooding. It is based on readily available information from major sources and is not a complete review of erosion, deposition and flooding in Great Britain. It is intended to be of use as background information and does not obviate the need for a full investigation of any particular site to establish its suitability for particular types of land use and development.

Executive Summary

Introduction

The operation of physical processes on hillslopes, within river channel networks and at the coast are of considerable interest to landowners, engineers, planners and land managers as well as the scientific community. Erosion processes operate to slowly break down the fabric of the land; the resulting material is then transported by wind or water and deposited elsewhere in new landforms. Land can be lost; channels and reservoirs choked with sediment leading to flood problems or reduced storage capacity; new land can be created; landforms can build up to give increased protection against erosion or flooding.

Erosion, deposition and flooding are natural phenomena and an integral part of the natural landscape, especially the dynamic environments of river channels and the coast. The processes can shape the landscape, forming, for example, the coastal cliffs and broad meandering rivers that are part of our natural heritage. They can create and sustain valued habitats and maintain important recreational beaches or sand dunes.

The processes only become hazards or problems when society encroaches into these dynamic environments either for housing or development on floodplains or coastal cliffs or for transportation and trade along canals, rivers and estuaries. Here, attitudes to erosion, deposition and flooding can vary dramatically. To some, the processes are an acceptable risk associated with living in desirable locations such as close to riverbanks or coastal cliffs; adjustments to the risk, such as floodproofing, can be made to mitigate against the effects of potentially damaging events. Others may be completely unaware of potential problems in an area; to them the sudden occurrence of an event may lead to unacceptable levels of loss.

In other instances the operation of the processes may go largely unnoticed by much of society. Erosion of riverbanks and hillslopes, for example, generally does not involve dramatic events; small amounts of material are regularly detached and carried away. The cumulative effects, however, can lead to serious consequences such as the gradual silting up of canals, rivers and estuaries which makes regular dredging essential to maintain navigable watercourses.

This study, commissioned by the Department of the Environment as part of its planning research programme, aims to provide an assessment of erosion, deposition and flooding processes, with particular reference to their significance to land use planning and development. The results have been presented as:

- a review of the occurrence and significance of the process (Rendel Geotechnics, 1995);

- a review of approaches to investigation and management of issues related to the processes, with special reference to the role of the planning system (this volume);

- a summary report, combining the key components of both the above volumes.

- a computerised database of records from over 1500 significant events over the last 200 years or so;

- a suite of 1:625,000 scale maps of;

  – the distribution of Records of Significant Events.

  – Potentially Vulnerable Areas.

Approaches to Investigation

The type and quantity of information needed to support decisions by planners and developers will vary according to the stage in the decision–making process. For both planners and developers the requirement for information ranges from a general awareness of the character of an area to site specific information. As a result a range of approaches are relevant of different stages in the planning and development process (Figure 1):

(i)   **General Assessment**; for many purposes, such as strategic policy formulation and preparing local plan proposals, a broad brush approach can contribute significantly to the safe, cost–effective development and use of land. General assessment of the principal physical and economic conditions in an area can provide a relatively quick appraisal through the collection and interpretation of readily available data sources. General assessment should be undertaken by the planning authority as part of its survey of the principal physical characteristics of its area when preparing or reviewing a development plan;

(ii)   **Site Reconnaissance**; both planners and developers have an interest in the selection of sites for development and thus may need to undertake appropriate site studies;

A **site review** is necessary to provide **planners** with as quick appraisal of potential sites for development and redevelopment, ensuring that, for example, they can be safely developed. Indeed, the nature of the "plan led" planning system dictates that development plan proposals should take full account of potential problems as failure to do so could lead to development in unsuitable locations. This level of investigation may also provide advance warning to **developers** of the potential difficulties associated with particular sites and an indication of the cost implications (i.e. **preliminary assessment**);

(iii)   **Detailed Assessment**; to demonstrate to the local planning authority that the proposed development is suitable and takes full account of all material planning considerations, including, as appropriate physical, economic and environmental factors. This level of investigation should enable developers to assess the level of risks associated with the development and the likely costs of special precautionary measures such as flood defence.

It is important to ensure that the effort spent by planners is appropriate to the level of investigation. In many instances it is not feasible to collect and collate all existing data relevant to any study, partly because some holders will not provide it and partly because considerable time may be needed to access widely dispersed data sets. Limitations to the time and resources available to planners will generally dictate that comprehensive geographical or thematic coverage is unlikely to be a realistic option. Investigations will, therefore, be a practical compromise addressing **key issues** that relate to specific policy needs, with effort directed towards areas where there is significant pressure for development.

The Requirement for Earth Science Information

In many instances planning and development decisions are influenced by social, economic and environmental factors. However, there are cases where the nature of the ground and the operation of physical will be relevant to the safe, cost–effective development of land. This is especially so on the coast, along river corridors and on some hillslopes, where the dynamic nature of the environment dictates that planners and developers need to be aware of physical factors such as erosion, depositions and flooding, mineral resources and conservation features.

Although planning considerations will not be uniform around the country, it is possible to identify a number of key issues that will need to be addressed. These include:

(i)   **Development**

●   the impact of erosion, deposition and flooding processes on development and the need for remedial and defence works;
●   the effects of development on the operation of physical processes

Figure 1 The sequence of investigations in the planning and development process.

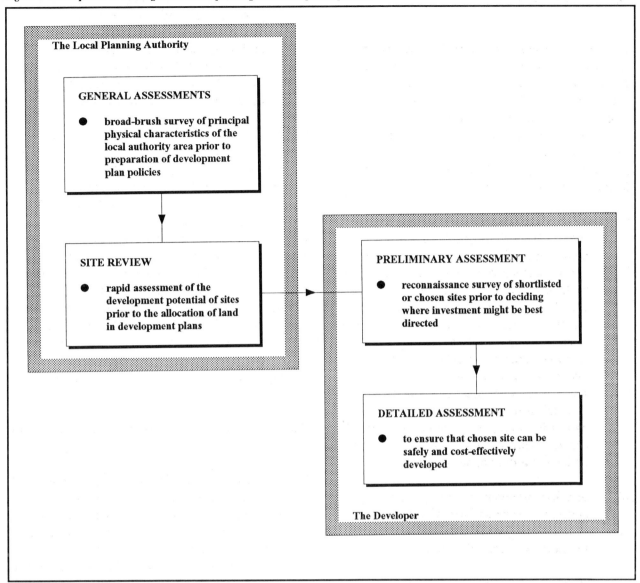

within a catchment or along the coast;
- the effects of development on mineral, water and conservation resources.

(ii) **Conservation**

- the importance of physical processes in creating and maintaining conservation features;
- the effects of conservation policies on diverting development into vulnerable areas away from conservation resources.

(iii) **Recreation**

- the importance of physical processes in creating and maintaining recreation resources such as beaches and sand dunes;
- the effects of recreational facilities on the stability of features such as sand dunes.

(iv) **Minerals**

- the occurrence and significance of mineral resources such as aggregates for construction and coastal defence, and building stone;
- the effects of mineral extraction on the operation of physical processes within a catchment or along the coast;

- the effects of mineral extraction on coastal, marine and riverine resources;
- the conservation opportunities provided by mineral workings.

These issues are not simply local concerns, with the influence of processes often extending well beyond a single authority area.

Key Considerations

The various earth science related issues can be addressed at a variety of scales in terms of three **key considerations**:

(i) the **risks** associated with hillslope, river and coastal processes (e.g. flooding, instability and erosion);

(ii) the **sediment budget** of a catchment or coastal system, which can be a major factor in the sustainability of conservation features and natural coastal defences;

(iii) the **sensitivity** of natural environment to natural or man–induced change. This, for example, can be an important factor in assessing the potential effects of development.

These considerations should be the focus of all investigations for planning purposes, from general assessments and site reconnaissance to detailed assessments, although their nature and significance will vary markedly between different environments.

Sources of Information

The information that is needed to address planning and management issues at different scales will generally be obtained in a variety of ways:

- Information specifically compiled by various organisations for other aspects of management, including monitoring programmes, catchment management plans, shoreline management studies, flood risk maps, estuary management plans, SSSI management plans, etc;

- Information previously collected as part of national mapping programmes or data collation exercises. Examples include British Geological Survey Geological Maps and the National Landslide Databank;

- Information that can be **derived** from readily available sources. Examples include the measurement of land loss through erosion from historical sequences of topographic maps or delineation of flood prone areas from detailed Ordnance Survey Maps;

- Information that is not currently available but will need to be collected as part of the investigation of specific sites or areas.

The information sources relevant at different levels of investigation are discussed in Chapters 5–7. However, it is important to stress that both planners and developers should try to make maximum use of the information held by the various bodies with an interest or responsibility in environmental management.

Identification of Vulnerable Areas

In general terms, potentially vulnerable areas can be readily identified from a range of available sources (Table 1). These sources can be used to define areas of the country where problems can be anticipated. However, it should be appreciated that the degree of hazard within these areas may vary according to the geomorphological setting. Although areas can be readily defined there are major problems in evaluating the **potential** for hazardous events to occur. Damaging events are often associated with very large return period storms (the Lynmouth floods of 1952 had an estimated return period of thousands of years) and are generally the product of a combination of circumstances (e.g. antecedent rainfall conditions, groundwater levels, land management practice).

Return period statistics are notoriously unreliable, especially when derived from very short periods of records as new events can lead to a significant modification in the calculated return period. For example, the Truro floods of January 1988 initially had an estimated return period of 350 years, using the procedures recommended in the Flood Studies

Table 1 Sources of information for the identification of vulnerable areas.

PROBLEM	ENGLAND AND WALES	SCOTLAND
Hillslope erosion	Soil Survey and Land Research Centre: - maps of erosion risk on agricultural land (1:250,000 scale) - maps of soil associations prone to water erosion can be <u>derived</u> from National Soil Maps (1:250,000 scale)	Macaulay Land Use Research Institute: - no published information on erosion prone soils
Wind erosion	Soil Survey and Land Research Centre: - maps of soil associations prone to wind erosion can be <u>derived</u> from National Soil Maps (1:250,000 scale)	Macaulay Land Use Research Institute - no published information on erosion prone soils
Flash floods	National Rivers Authority: - areas prone to flooding shown on S.105 survey maps (various scales)	Various possible sources, including: - Island and Regional Councils - River Purification Boards
Channel Instability	No systematic collection of records. Channel positions on present day topographical maps can be compared with positions on historical maps.	
Lowland flooding	National Rivers Authority: - areas prone to flooding shown on S.105 survey maps (various scales)	Various possible sources, including: - Island and Regional Councils - River Purification Boards Floodplains extent can be <u>derived</u> from Macaulay Land Use Research Institute soil maps (1:250,000 scale) or British Geological Survey geological maps (1:50,000/1:63,360 scale)
River and estuary sedimentation	No systematic collection of records. Channel depth on present day Admiralty Charts can be compared with positions on historical charts.	
Coastal flooding	National Rivers Authority: - areas prone to flooding shown on S.105 survey maps (various scales)	Various possible sources, including: - Island and Regional Councils - River Purification Boards Flood prone area can be <u>derived</u> from Macaulay Land Use Research Institute soil map (1:250,000 scale) or British Geological Survey geological maps (1:50,000/1:63,360 scale). Ordnance Survey maps can be used to identify the extent of land below 5m.
Coastal cliff erosion	Distribution of recorded landslides can be obtained from the National Landslide Databank (Rendel Geotechnics). No systematic collection of erosion records. Cliff positions on present day topographical maps can be compared with positions on historical maps.	
Mobile sand dunes	Soil Survey and Local Research Centre: - maps of soil associations prone to wind erosion can be <u>derived</u> from National Soil Maps (1:25,000 scale)	Macaulay Land Use Research Institute: - no published information on erosion prone soil Scottish Natural Heritage: - The Beaches of Scotland (Ritchie and Mather, 1985)

Report (NERC, 1975); a reappraisal of the risk after further floods in October 1988 suggested a return period of 50 years for the January event.

Perhaps the single most important limitation on hazard assessments is the limited data sets of, for example, rainfall or flow records from which predications have to be made. Benson (1960) demonstrated that to achieve 95% reliability on the estimate of discharge for a 50 year flood event required 110 years of records. Such lengthy data sets are not common in Great Britain. A different perspective on hazard potential is provided by considering the degree of risk in terms of the **standard of protection** which exists in an area. MAFF, for example, have recently set out **indicative standards of protection** in terms of flood return periods for five subjectively expressed current land use bands.

Notable difficulties also arise in the prediction of erosion rates, especially on coastal cliffs. Cliff recession is an episodic process; there may be little or no erosion for a long period, followed by sudden rapid recession in response to a large storm. The "rate" of recession may, therefore, depend on the timescale chosen. Furthermore, the main causal factors such as wave undercutting are inherently unpredictable. This uncertainty creates problems for coastal managers when attempting to define the degree of risk to cliff top property, either for identification of "set-back" lines by planners or the choice of **erosion control technique** by engineers.

Approaches to Management

Voluntary acceptance of some level of risk is inherent in every decision to purchase property, carry out economic activity or provide services in vulnerable areas. The resulting costs that can arise when damaging events occur are borne by the affected parties through **maintenance and repair, emergency action** or offset by **insurance** claims or **legal action**. Frequently the acceptance of risk makes sound economic sense. Indeed, in many instances the benefits of placing property, infrastructure and services in vulnerable areas may outweigh the risks from erosion, deposition and flooding.

Harbour and navigation authorities need to carry out maintenance works to ensure that navigable channels remain open for commercial or leisure vessels. Dredging is the most important activity. Although operations are, in general, authorised by statutory provisions they can be constrained by the need to find safe disposal sites on land or at sea. Problems are enhanced when the dredged sediment is contaminated by heavy metals or other potentially harmful substances that, in the past, had been freely discharged in rivers or coastal waters.

Erosion, deposition and flooding present significant **hazards** to land use and development, whilst having important roles in the creating or maintaining valuable **resources**. These factors have become increasingly difficult to separate. Indeed, the complexity of the inter-relationships between the processes and management practices, and their combined effects on the variety of land use interests highlight the need for a **coordinated approach** to decision making. This would generally involve **regional** awareness of how the effects of an activity or proposed development at one location can be transmitted through a catchment or over a coastal system and create a variety of problems to local amenity or conservation interests, as well as increasing the level or risk to other developments.

The contrast between the broad scale operation of **catchments** or **coastal systems** and the limited extent of most local authority areas (Scotland's administrative regions are a notable exception) can lead to management problems, with local authorities concentrating on local issues rather than taking a strategic view of how development can cause problems for their neighbours. Such problems can be overcome through the preparation and use of management plans for catchments and coastal systems which address a range of issues from land use and conservation to flood and coastal defence.

A wide variety of responses are available for managing physical processes on hillslopes, rives and the coast (Table 2). The most appropriate option will depend on the nature of the problem, the level of acceptable risk, the availability of resources and the statutory powers available to interested bodies or authorities to tackle the problems. In most cases the response will be complex, involving a variety of measures adopted by different organisations at different locations within a catchment or coastal system.

The scale and complexity of erosion, deposition and flooding issues dictate that conventional engineering solutions cannot **prevent** the occurrence of potentially damaging events. More realistic aims would involve **reducing** the frequency of damaging events and **minimising** their impact, through:

- providing reliable early warning systems;
- providing appropriate levels of protection to vulnerable communities;
- minimising the effects of human disturbance on the occurrence of erosion, deposition and flood events;
- reducing the damage potential by discouraging inappropriate development in vulnerable areas.

The Role of the Planning System

The planning system can be used in a variety of ways to achieve management objectives, including:

- ensuring that certain river channel or foreshore maintenance works do not have an adverse effect on environmental interests (Chapter 8);
- ensuring that new development does not restrict access to watercourses for channel maintenance works (Chapter 8);
- avoiding locating new development in unsuitable areas or specifying restrictions on housing occupancy in risk areas (Chapter 9);
- discontinuing or restricting existing land uses or buildings in vulnerable areas through discontinuance orders or compulsory purchase and providing compensation to affected parties (Chapter 9);
- ensuring that precautions are taken to prevent runoff from new developments increasing flood risk (Chapter 10);
- ensuring that development does not adversely affect floodplain storage and, hence, increase flood risk (Chapter 10);
- ensuring that development does not affect coastal cliff stability or lead to an increase in coastal erosion (Chapter 10);

- ensuring that coast protection works and new flood defence works are compatible with land use planning and conservation objectives in an area (Chapter 11).
- supporting environmental management objectives by ensuring that development does not lead to a decline in value of conservation sites dependant on continued operation of physical process (Chapter 12).

There are considerable opportunities to prevent or reduce damage to new development through the consideration of erosion, deposition and flooding at all stages of the planning framework from regional to local levels. Local planning authorities should also liaise with catchment and shoreline managers in developing a coordinated approach to managing erosion, deposition and flood related issues.

(i) **Regional Planning**; Natural processes operate over a much broader scale than individual local authority areas. As a result land use policies and proposals in one area may have an adverse effect downstream or on neighbouring sections of the coast. Regional planning groups should, therefore, liaise with catchment managers (e.g. the NRA in England and Wales) and Coastal

Table 2 Management responses.

Acceptance of the risk
• maintenance and repair
• emergency planning
• insurance
• litigation
Avoiding Vulnerable Areas
• land use planning
• managed retreat
Reducing the Likelihood of Potentially Damaging Events
• hillslope management
• runoff management
• floodplain management
• river channel management
• coastal cliff management
• foreshore management
Protecting Against Potentially Damaging Events
• early warning
• defence schemes
• building modifications

Defence Groups to identify those erosion, deposition and flood related issues that need to be considered within the context of a catchment or coastal system. Possible examples include the effect of development on the disruption of the natural movement of coastal sediments and the reduction of floodplain storage within catchments.

(ii) **Development Plans; Structure Plans** or **Unitary Development Plans (Part I)** provide an excellent opportunity for identifying the extent of erosion, deposition and flooding problems in an area and outlining the range of policies that are to be adopted in these areas. The scale of provision to be made for housing, employment opportunities, etc. in the area as a whole and the breakdown for each district should take account of any erosion, deposition and flood related issues, including the need for flood and coastal defences.

The **Local Development Plan** or **Unitary Development Plan (Part II)** can be used to set out detailed policies to address these issues, the basis for determining planning applications and the types of planning conditions normally expected to be met. In identifying suitable sites for development or redevelopment, the local planning authority should consider whether any necessary flood or coastal defences can be undertaken without adversely affecting other interests. Local Plans should define those areas where particular consideration of the implications of erosion, deposition and flooding is necessary, either in narrative form or by reference to maps of resources and constraints.

(iii) **Planning Applications;** The handling of planning applications for development in areas prone to erosion, deposition and flooding needs to take full account of the potential problems that these processes could cause to the proposed development, the neighbouring area and elsewhere within a catchment or coastal system. The authority should then determine whether a proposed development should proceed, taking into account all relevant material considerations of which natural hazards are

only one. A development may be approved subject to **conditions** specifying measures to be carried out in order to minimise risk or reduce the effects of the development to other interests. These may include:

- building design; minimum floor heights or means of escape in flood risk areas, flexible constructions on unstable coastal slopes;
- occupancy; limiting periods of occupancy in flood risk areas to periods when flood risk is significantly reduced or controlling the use of basements for apartments in vulnerable areas.

(iv) **Environmental Assessment;** The UK is bound by **EC Directive 85/337/EEC** on "the assessment of the effects of certain public and private projects on the environment". This directive requires an **environmental assessment** (EA) to be carried out before development consent is granted for certain types of major project, listed in two Annexes to the directive.

The planning system is one of the main instruments for taking account of an EA under the **Town and Country Planning (Assessment of Environmental Effects) Regulations 1988**. The regulations apply to certain projects that require planning permission and included within the Annex II schedule are: canalization or flood relief works; dams and reservoirs and coast protection works. In deciding whether an EA is required, local planning authorities should bear in mind the potential effects of these projects on the level of risks elsewhere, local amenities and conservation interests. EA may be particularly important for projects:

- in particularly sensitive or vulnerable locations;
- which could give rise to particularly complex adverse effects, for example, in terms of disruption in the transport of coastal sediment or changes in the flow regime of a river.

(v) **Liaison with Catchment and Shoreline Managers**; In recent years there has been a growing concern about the traditional response to the threats posed by natural hazards. Flood and coastal defence works have been very successful in protecting communities from the threat of erosion and flooding. However, construction of the defences has had a range of effects on the physical environment. River and coastal engineering is only one of a range of policy options available. The choice of option should depend on the nature of the catchment or coastline, land use policies, conservation needs, benefits, costs and resources.

Decisions to proceed with engineering works need to be made within the framework of a range of **land use planning** policies. Bearing in mind that there is essentially a presumption in favour of developments that are in accordance with the development plan it is important that development plan policies and proposals take into account flood and coastal defence issues. Where sites are at risk from erosion, deposition or flooding it is important to ensure that a satisfactory defence infrastructure exists and can be maintained over the life of the development, **or** that new defences can be provided by the developer that do not affect other interests.

The Role of the Developer

The responsibility for determining whether land is suitable for a proposed development lies with the developer and/or the landowner. Obviously it is in a developer's own interests to determine whether a site is in a vulnerable location as any future problems will affect the value of the site and both its development costs and maintenance. If there are any reasons for suspecting potential problems the developer should instigate appropriate investigations to determine whether:

- the land is capable of supporting the load to be imposed;
- the development will be threatened by erosion, deposition or flooding problems;

- the development will affect the level of risk from erosion, deposition and flooding problems to property, infrastructure and services in adjacent areas or elsewhere within a catchment or coastal system.

Although the majority of questions facing developers relate to site specific issues, the latter point highlights the need to be aware of how the effects of development at a particular location can be transmitted over a wider area. Here, an understanding of how processes operate within large physical systems (catchments and coastal systems) is necessary.

The assessment of erosion, deposition and flooding problems and the associated risk requires careful professional judgment. Developers should seek expert advice about the likely consequences of proposed developments on the physical environment. This advice will generally involve some form of **investigation** into the nature of the problem and should provide an indication as to whether the site is suitable or whether precautionary measures are needed to prevent problems affecting the site or neighbouring areas.

The developer should provide a **Site Report** containing sufficient information on erosion, deposition and flood matters to enable the local authority to review the planning application. Indeed, the authority is entitled to require the developer to seek suitable expert advice. It is important to stress that the developer needs to investigate not only the conditions of the proposed site, but also whether the development could adversely affect the surrounding land (e.g. as a result of accidental water leakage or removal or support) or elsewhere in the catchment or coastal system (e.g. by disrupting the natural transport of sediment around the coast).

Future Research Needs

Erosion, deposition and flooding can present significant problems for land use planning and development, most notably through:

- the impact of the processes on property, infrastructure and services;
- the effects of development on the degree of risk elsewhere;

- the conflicts generated by the selection of hazard management strategies.

In areas where erosion, deposition and flooding are likely to impose constraints to development and land use, decision makers will need to consider identifying those areas where particular consideration should be given to these issues. This requires access to reliable technical information and advice. An equally pressing need is for local planning authorities to be aware of the ways in which the planning system can be used to achieve management objectives. In the past this potential has seldom been realised. There is, therefore, a clear need to improve the awareness of the role of the planning system as an instrument in the management of erosion, deposition and flooding issues. This could involve considering the following recommendations:

- reviewing the use of discontinuance orders or compulsory purchase in managing risks or compulsory purchase in managing risks along river corridors and in the coastal zone;
- extending the scope of the Building Regulations to cover flooding issues;
- preparing planning guidance to address catchment planning issues;
- monitoring the effectiveness of flood risk maps and management plans as a source of information for the planning system;
- raising awareness of the Applied Earth Science Mapping programme;
- commissioning demonstration projects to identify approaches to incorporating flood potential, sediment budget and sensitivity information in the planning system.

Contents

List of Figures

List of Tables

List of Abbreviations

ADAS	Agricultural Development and Advisory Service
AoD	Above Ordnance Datum
AONB	Area of Outstanding Natural Beauty
BGS	British Geological Survey
BSI	British Standards Institution
CCS	Countryside Commission for Scotland
CCW	Countryside Council for Wales
CEC	Crown Estate Commissioners
CMP	Catchment Management Plans
DCPN	Development Control Policy Note
DoE	Department of the Environment
DNH	Department of National Heritage
DSS	Department of Social Services
DTp	Department of Transport
EA	Environmental Assessment
EC	European Community
EN	English Nature
ES	Environmental Statement
EBA	Environmentally Sensitive Area
GCR	Geological Conservation Review
GDO	Town and Country Planning General Development Order 1988
GVP	Government View Procedure
HWM	High Water Mark
ICE	Institution of Civil Engineers
IDB	Internal Drainage Board
IWEM	Institution of Water and Environmental Management
JNCC	Joint Nature Conservation Committee
LPA	Local Planning Authority
LWM	Low Water Mark
MACC	Military Aid to Civil Communities
MAFF	Ministry of Agriculture, Fisheries and Food
MLWS	Mean Low Water, spring tides
MPG	Minerals Planning Guidance note
NERC	Natural Environment Research Council
NCC	Nature Conservancy Council
NCR	Nature Conservation Review
NRA	National Rivers Authority
PDO	Potentially Damaging Operation
POL	Proudman Oceanographic Laboratory
PPG	Planning Policy Guidance note
RIGS	Regionally Important Geological Site

RPA	River Purification Authority
RPB	River Purification Board
RPG	Regional Planning Guidance note
RSPB	Royal Society for the Protection of Birds
RTPI	Royal Town Planning Institute
SCOPAC	Standing Conference on Problems Associated with the Coastline
SDD	Scottish Development Department
SMP	Shoreline Management Plan
SOAFD	Scottish Office Agriculture and Fisheries Department
SNH	Scottish Natural Heritage
SPA	Special Protection Area
SSSI	Site of Special Scientific Interest
STWS	Storm Tide Warning Service
UDP	Unitary Development Plan
UK	United Kingdom
WO	Welsh Office
WRVS	Women's Royal Voluntary Service
WWF	World Wide Fund for Nature

1 Introduction

Background

The Department of the Environment (DoE) undertakes geological and earth-science related research as part of its Planning Research Programme. The programme aims to provide information on planning policies, planning processes and the context within which the planning system operates. One of the areas of research is directed towards minerals and land instability (DoE, 1994).

Land which is actually or potentially unstable is widespread in Great Britain. Some instability arises from natural processes. Often it is caused or accelerated by human activities. Problems include landslides and rockfall associated with natural slopes and cliffs, or with embankments, cuttings, quarry faces or spoil tips. Subsidence of the surface is associated with mined ground, natural caves and fissures, solution of salt, or underground combustion of coal. Open, or poorly covered, mine entries may be dangerous. Foundations may be damaged by compression of sediments or of landfill, or by swelling and shrinking of clays or other materials. Britain even experiences damaging earthquakes but, fortunately, these are rare.

In addition, damage can arise from surface flooding from rivers or the sea. Erosion can exacerbate ground movements or cause loss of soils. Deposited sediments can block waterways and pipes leading to flood problems. Rising groundwater can cause problems for tunnels and basements. Natural chemical substances within the ground may react with buildings materials. Some of these are poisonous to people or livestock. Emissions of natural gases can be explosive, suffocating, poisonous or radioactive.

The costs of these problems in damage to property, injuries and deaths are poorly known but are undoubtedly very large. Many problems can be avoided or reduced by proper precautions. These include planning to avoid inappropriate development in areas most at risk, proper site investigations, appropriate precautionary and remedial works, and design of structures to, for example, accommodate ground movements or exclude hazardous gases. But this is only possible if the problems are appreciated. This requires adequate information for planning of land use, control of development and application of the Building Regulations. Such information can also guide strategies for reclamation of derelict land.

Research is undertaken to establish the nature and extent of problems and the best technical and administrative approaches for dealing with them. The results, thus, contribute to the objectives of the Environment White Paper "This Common Inheritance" of securing a physically safe environment, of minimising risks to human health and the environment, and recycling derelict and contaminated sites (DoE, 1990).

A series of **National Reviews** of specific ground-related problems have been commissioned by the Department:

- Review of Landsliding (Geomorphological Services Ltd 1986-87);

- Review of Mining Instability (Arup Geotechnics, 1991);

- Preliminary Assessment of Seismic Risk (Ove Arup and Partners, 1993);

- Review of Natural Underground Cavities (Applied Geology (Central) Ltd, 1994);

- Review of Foundation Conditions (Wimpey Environmental Ltd, 1994);

- Review of Natural Contamination (British Geological Survey, 1994).

This report presents the results of the DoE research contract PECD 7/1/412 entitled "Review of Erosion, Deposition and Flooding in Great Britain". It presents a general review of the investigation and management of issues associated with these three natural processes and their impact on the development and use of land (Table 1.1).

Programme of Work

A **desk study** review of hazard mitigation and management approaches was undertaken. This involved specific reviews of:

- methods of investigating and monitoring erosion, deposition and flooding;

- measures for hazard reduction, including planning and prevention and remedial techniques, based mainly on British experience and practice but also taking account of overseas practices where relevant;

- methods of risk assessment, paying particular attention to the significance of the hazards in constraining present and potential future land uses.

The review of the planning system addressed the following topics:

- the administrative and legislative framework for dealing with problems and, in particular, considerations which need to be taken account of in preparation of planning policies and proposals, planning and Building Regulations decisions, or which relate to permitted development rights under the General Development Order;

- the current availability and reliability of information needed for formulating planning policies and deciding planning applications, identifying where data is held and in what format;

- methods of summarising and presenting information so that it can be taken into account readily in the planning and development processes.

The work was based on: discussions with local authorities; contacts with NRA regional offices; discussions with officers from River Purification Boards in Scotland; selected case studies; and a review of the administrative and legislative framework. During 1992 a sample of local planning authorities in the South East, Yorkshire and Humberside, central and eastern Scotland and Wales were contacted. The main aims were:

- to identify the occurrence and significance in land use terms, of notable examples of erosion, deposition and flooding;

- to identify the planning response to such events;

- to consider liaison arrangements with the NRA; and

- to identify relevant development plan policies.

Discussions were also held with key personnel from a limited number of local planning authorities throughout Great Britain concentrating on: local planning responses to erosion, deposition and flooding; the use of technical information in support of decision making and relationships with other authorities and bodies with hazard management responsibilities.

Each of the National Rivers Authority (NRA) Regions in England and Wales has been contacted. The principal topics covered have been: planning liaison arrangements; the use of NRA guidance to local planning authorities and identifying what data is held on flooding and erosion and the extent to which this data is used by local planning authorities.

In addition to the above a series of five brief **case studies** have been undertaken. These were located in:

- Tayside;
- Humberside;
- Solent;
- Thames Region of the NRA; and
- River Dee, North Wales.

The case studies were chosen to cover a range of planning and management responses to problems in different environments (Table 1.2). They were intended to illustrate:

Table 1.1 The aims and objectives of the study.

The **aim** of the study has been to provide an assessment of the nature of erosion, deposition and flooding process in Great Britain, and an understanding of their significance for development and the use of land, and for the operation of the planning system.

The **objectives** of the study were to review:

a)   the types, causes and extent of significant occurrences of erosion, deposition and flooding in Great Britain;

b)   problems resulting from the interactions of these phenomena and their relationships to types of land instability and changes in land uses;

c)   their significance for conservation and maintenance of natural habitats and systems;

d)   methods for investigating and monitoring erosion, deposition and flooding;

e)   measures for hazard reduction including planning, preventive and remedial techniques;

f)   the current availability and reliability of information needed for formulating planning policies and deciding planning applications;

g)   methods of risk assessment, including the significance of flooding in relation to existing uses of land and land use potential, and the definition of areas in which development is constrained by erosion, deposition or flooding;

h)   methods of summarising and presenting information so that it can be taken into account readily in the planning and development process;

i)   the administrative and legislative framework for dealing with the problems;

j)   gaps in existing knowledge, and to set out priorities for further investigations; and

k)   to prepare a draft framework of advice to planners and developers.

The work has been organised in four related Tasks, as follows:

Task 1:   Occurrence and Significance of Erosion, Deposition and Flooding in Great Britain.
Task 2:   Review of British and overseas practices for investigation, management and risk assessment.
Task 3:   Review of the planning system responses to hazards, and improvements to practice.
Task 4:   National and summary reports and dissemination of results.

This draft report addresses Tasks 2 and 3: the investigation and management of erosion, deposition and flooding with special reference to the role of the planning system.

- the use of data on erosion, deposition and flood events by local planning authorities;

- the role of central government in providing advice and guidance;

- the type of data used;

- legal/administrative issues;

- whether conflicts have arisen between bodies responsible for managing the processes.

Descriptions of each case study are presented in the **Methodology Report** (Rendel Geotechnics, 1995b).

Contents of this Report

This Report has been sub–divided into 4 distinct Parts. Part I provides an introduction to the nature of erosion, deposition and flooding issues in Great Britain, and is intended to act as a link with the complementary report "The Occurrence and Significance" (Rendel Geotechnics, 1995a). A brief overview of the legislative and administrative framework relevant to these issues is presented in Chapter 3. In Part II a general model of investigation is presented, involving a general discussion (Chapter 4) followed by the information needed for general assessment (Chapter 5), site reconnaissance (Chapter 6) and detailed assessment (Chapter 7). The range of available management options are outlined in Part III including aspects of

Table 1.2 Description of case studies (details are presented in Rendel Geotechnics, 1995b).

CASE STUDY	COMMENT
The Solent	This case study considered planning responses to forecast increases in sea levels which could result in increased coastal erosion and flooding. By considering two authorities, comparisons were made between urban and rural areas. In addition the study was able to consider planning concerns resulting from intense recreation and development pressures.
Humber Estuary	The majority of this study area is rural with most development pressure focused on the west side of Kingston-upon-Hull. Extensive areas of both Boroughs are susceptible to estuarine and river flooding and the NRA is in the process of upgrading flood defences.
Tayside	This case study specifically considered planning responses to river flooding following flood events in 1990 and 1993. Particular consideration was given to responses to the flooding of part of Perth in January 1993. The location of this case study in Scotland allows comparisons to be drawn between institutional arrangements in Scotland compared to the remainder of Great Britain.
NRA Thames Region	Three brief studies were carried out in different parts of the Thames Region of the NRA each focusing on a specific planning issue: – the relationship between the draft Blackwater Catchment Management Plan prepared by the NRA and local planning authority activities; – strategic planning provisions for major housing development in the Kennet Valley and the consideration of flood risk in the development of these proposals; and – different planning authority responses to the construction of major flood relief works for Maidenhead, Windsor and Eton.
Dee Meanders	This case study focused on the proposal by the conservation agencies to make an area of meandering river an SSSI, and the NRA's objection on the grounds that this could affect their ability to carry out flood defence maintenance.

hazard management (Chapters 8 –11) and the management of conservation sites and areas (Chapter 12). Part IV addresses the role of the planning system in the investigation and management of erosion, deposition and flooding (Chapter 13), setting out a framework of advice for planners and developers (Chapter 14) and conclusions and recommendations (Chapter 15).

Chapter 1: References

Applied Geology (Central) Limited, 1994. Review of Natural Underground Cavities. Report to DoE.

Arup Geotechnics, 1991. Review of Mining Instability. Reports to DoE.

British Geological Survey, 1994. Review of Natural Contamination. Reports to DoE.

Department of the Environment, 1990. This Common Inheritance. Britain's Environmental Strategy. HMSO.

Department of the Environment, 1994. Geological and Minerals Planning Research Programme. Annual Report for 1993/94.

Geomorphological Services Limited, 1986–1987. Review of Landsliding. Reports to DoE.

Ove Arup and Partners, 1993. Preliminary Assessment of Seismic Rick. Reports to DoE.

Rendel Geotechnics 1995a. The Occurrence and Significance of Erosion, Deposition and Flooding in Great Britain.

Rendel Geotechnics 1995b. Erosion, Deposition and Flooding in Great Britain. Methodology Report. Open File Report held at the DoE.

Wimpey Environmental Limited, 1994. Review of Foundation Conditions. Reports to DoE.

2 Background to Managing Erosion, Deposition and Flooding Problems in Great Britain

Introduction

In many instances planning and development decisions are influenced by social, economic and environmental factors. However, there are cases where the nature of the ground and the operation of physical processes will be relevant to the safe, cost–effective development of land. This is especially so on the coast, along river corridors and on some hillslopes where the dynamic nature of the environment dictates that planners and developers need to be aware of physical factors such as erosion, deposition and flooding, and their interrelationships with resources, and conservation and recreation features.

Although considerations will not be uniform around the country, it is possible to identify a number of **key issues** that will need to be addressed in most areas. These include:

(i)   **Development**

● the impact of erosion, deposition and flooding processes on development and the need for remedial and defence works;

● the effects of development on the operation of physical processes within a catchment or along the coast;

● the effects of development on mineral and conservation resources;

(ii)   **Conservation**

● the importance of physical processes in creating and maintaining conservation features;

● the effects of conservation policies on diverting development into vulnerable areas away from conservation resources;

(iii)   **Recreation**

● the importance of physical processes in creating and maintaining recreation resources such as beaches and sand dunes;

● the effects of recreational facilities on the stability of features such as sand dunes;

(iv)   **Minerals**

● the significance of erosion, deposition and flooding to sustaining mineral resources such as aggregates for construction and coastal defence, and building stone;

● the effects of mineral extraction on the operation of physical processes, within a catchment or along the coast;

● the effects of mineral extraction on coastal, marine and riverine resources;

● the conservation opportunities provided by mineral workings.

The range and variety of these issues serve to emphasise that erosion, deposition and flooding should not be viewed only as **hazards** or **risks** to society. Although they may result in considerable losses, the management of these processes should not be directed solely towards controlling their operation as this may have significant consequences for other environmental interests. Indeed, the potential for conflict between the authorities responsible for flood and coastal defence with conservation agencies or other environmental interests is a recurrent theme throughout this report. Reconciling these potential conflicts is an important factor in ensuring that development, in catchments and on the coast, is **sustainable** (i.e, development meets the needs of the present without compromising the ability of future generations to meet their own needs; Brundtland Commission, 1987).

The Nature of the Problems

Erosion, deposition and flooding are natural phenomena associated with hillslopes, river systems and the coast. The **hillslopes** are the site of soil erosion and deposition by water and, occasionally, wind; slope failure through landslides and debris flows also occur (these have been investigated previously in the Review of Landsliding in Great Britain; Geomorphological Services Limited, 1986–87). Both landslides and erosion supply sediment to the **river networks** where channel migration, flooding and deposition are the principal forms of geomorphological activity. The rivers reach the coastline in their **tidal estuaries** where sedimentation, flooding and erosion of soft materials by tidal currents are the dominant processes. On the **open coast** there are complex patterns of erosion and deposition, with tidal flooding in low lying areas.

These processes can become significant problems to society as land use and development encroaches into vulnerable areas; settlements with their associated transport infrastructure and services have expanded into areas where erosion, deposition and flooding have always existed. This is readily apparent in the coastal lowlands of eastern England on the broad floodplains of Britain's major rivers and on the soft rock cliffs of eastern and southern England. At the same time, human occupancy of the landscape has transformed the intensity and frequency of many of these processes, heightening problems in some areas. For example,

- **agricultural practice**; the general post–war intensification of agricultural production has been accompanied by removal of hedgerows, use of heavy vehicles, monoculture and winter cereal cultivation. All are believed to have contributed to an increase in soil erosion. Land drainage has had a significant effect on the speed at which both ground and surface water is carried to stream channels, increasing flood risk;

- **forestry**; after World War II timber production was seen to be of strategic importance, with the result that large areas of upland Britain have been afforested in the last 50 years or so. Between 1945 and 1983, 700,000ha were planted by the Forestry Commission, representing the single largest land use change in Britain over this period (Acreman, 1985; Best 1976). About 60% of this land has been ploughed prior to planting (Taylor, 1970). It is believed that this has led to temporary increases in hillslope erosion and supply to sediment to stream channels, and more permanent changes in flood behaviour further down a catchment;

- **development**; the increasing demand for housing and employment opportunities by a growing population has led to an increased utilisation of floodplains and cliff–top locations. Frequently and despite the risks, these are viewed as desirable settings for expensive housing. Development has often been accompanied by: an increase in the amount of impermeable surfaces within a catchment, increasing and accelerating runoff; reduction in floodplain storage following flood defence works; changes in flood behaviour following the construction of sewage and stormwater drainage systems; uncontrolled surface water discharge into slopes leading to increased erosion problems;

- **river channelisation and flood defences** have often led to modifications to the patterns of erosion and deposition along a stretch of river, occasionally leading to river channel migration, and, in places, an increase in flood risk downstream. Flood embankments can prevent the natural overbank deposition of suspended sediments, leading to enhanced sedimentation downstream;

- **coastal defences** have frequently resulted in a disruption in the supply and transport of sediment around the coast. This can lead to increased coastal erosion or flood risk, especially where natural defences such as sand dunes or beaches are deprived or a regular supply of sediment;

- **dredging operations** in some rivers and estuaries can lead to the movement of further sediment back into the deepened channel, resulting in a need for repeated dredging. This problem was recognised in the Thames estuary where the removal of $3Mm^3$ of dredged sediment each year to the outer estuary merely resulted in its immediate transfer back into the dredged channel, (Inglis and Allen, 1957). By disposing of the spoil at sea, away from the outer estuary, the annual amount of dredging was reduced to around 0.25Mm^3;

- **water supply**; the supply of cheap and clean water for the lowland centres of population from upland areas has transformed the behaviour of many rivers. Reservoir storage is essential to ensure a regular supply; these structures have also helped regulate flood flows. However, there have also been downstream implications for river channel form and habitats;

- **conservation**; the growing awareness of environmental issues since the 1940s has led to the designation of green belts, valued landscapes (e.g. National Parks, AONB's, National Scenic Areas) and conservation sites (e.g. SSSIs, National Nature Reserves, etc.). Development has been largely directed away from these areas, often increasing the pressure for urban expansion in floodplain areas or low lying coastal areas, e.g. in the Thames Valley. Conservation status has, of course, been used to restrict development in some areas of potential problems; it may, therefore, be the best land use option in vulnerable areas.

The problems associated with these processes are generally similar: loss of agricultural productivity, damage to property, services and infrastructure, the need for regular maintenance to ensure unhindered use of waterways; and occasionally death or injury. It has been estimated that the average level of damage, maintenance and defence costs associated with these processes probably exceeds £300M per year (Rendel Geotechnics, 1995); these costs are, of course, spread through many levels of the economy, from individuals to industry, local authorities to national government and include:

(i)   **direct damages** caused by the effects of erosion and deposition or the physical contact of floodwater with properties and their contents;

(ii)   **indirect damages** arising as a consequence of direct damage, including: traffic disruption, loss of production, evacuation costs, etc.;

(iii)   **intangible damages** ranging from anxiety and stress to ill health related to the general inconvenience caused by the event.

However, it is the geographical extent and the intensity of damage, disruption or personal losses that set some processes apart. Flooding – flash floods in upland areas, lowland river floods and tidal floods – is the most dramatic and costly problem for society. Examples of particularly distressing and costly events can be found in the historical record for many parts of Britain; some of the worst include:

- the flash floods in the Lynmouth area of Devon on 15 August 1952 when 34 were killed with £9M of damage (at 1952 prices);

- the lowland floods of March 1947 which affected rivers throughout South Wales and much of England. The resulting damages were probably in excess of £500M at current prices;

- the east coast floods of January and February 1953 when over 300 died and damages were an estimated 900M (at current prices);

- the Severnside coastal floods of 1606 when about 2000 people drowned as sea defences were overtopped.

By contrast with the spatially extensive problems associated with flooding, other processes tend to create **site specific** or localised difficulties. Even so, they can still pose a significant threat to construction and development or lead to high maintenance costs to alleviate the effects of the processes. The erosion of coastal cliffs and sedimentation in rivers or estuaries can be a significant constraint to human activity, as illustrated by:

- the major coastal landslide at Holbeck Hall, Scarborough in June 1993 which is likely to have resulted in excess of £3M of damage and repair works;

- the loss of 75Mm^3 (800 ha) of land from the rapidly eroding Holderness cliffs over the last 100 years;

- the annual maintenance dredging costs in excess of £1M incurred at Harwich and Liverpool. At Kings Lynn, the approaches

7

have to be resurveyed every two weeks with navigation buoys repositioned up to 100 times a year.

Deposition within river channels and canals can lead to serious maintenance and operational problems. In 1993, for example, British Waterways spent over £3M on dredging, involving the removal of 300,000 tonnes of material. These operations are necessary to ensure that British Waterways fulfils its statutory obligations, but the need to dispose of the material on land can lead to conflict with environmental interests.

Other processes such as hillslope erosion, wind erosion and channel migration can lead to notable problems for affected landowners and can lead to difficulties where infrastructure and services cross vulnerable areas. The implications of these problems are easy to dismiss as trivial; the following examples should serve to demonstrate that they can lead to serious problems:

- soil erosion and mudflood problems in the South Downs during October 1987 probably resulted in £0.75M of damage, especially in and around Rottingdean;

- channel scour and erosion around the piers of a railway bridge at Glanrhyd, Dyfed led to the bridge collapsing under the weight of a train in October 1987; four people died;

- the Culbin Sands disaster of 1694 and following years led to over 20–30km² of fertile farmlands, near Findhorn on the Moray Firth, being buried by up to 30m of loose sand. The estimated damages were probably the equivalent of £25M, at present prices.

Many relatively minor erosion events can also achieve importance because they supply sediment to rivers or the coastal zone. Sediment supply occurs in areas of **hillslope erosion** and where **river channel migration** cuts through areas of stored sediments resulting from past phases of erosion under different climatic conditions or from an extreme flood in the recent past (e.g. spreads of glacial deposits or floodplain alluvium). Once in the channel, the sediment size is important in determining how far it is carried before being temporarily stored in features such as point bars or as spreads on the river bed. In short rivers the suspended load may reach the estuary in a single flood, but coarser sediments may become incorporated in the floodplain. Deposition within

the channel can, of course, reduce its capacity and lead to flood problems, as reported for the River Spey in Grampian and the Findhorn in Highland. It can also lead to navigability problems; scouring of bridges and other engineering structures; sedimentation and loss of reservoir capacity; a decline in water quality. Deposition around the coast can help maintain natural coastal defences, such as sand dunes and beaches, which protect vulnerable areas from flooding or cliff erosion.

The Significance for the Environment

The processes of erosion, deposition and flooding can also be of benefit to society, notably through shaping, maintaining or creating nationally and internationally valued landscapes, habitats or geological features. In broad terms the processes can:

- **maintain** habitats in river environments and on the coast, through regular inundation or supply of sediment;

- **maintain** geological exposures or valued landforms along the coastline through continued erosion;

- **preparing** gravel bed rivers for spawning fish such as salmon;

- **create** valued landforms such as the fluvial geomorphological features associated with flash flooding or channel migration;

- **stimulate change** through promoting instability, ensuring that habitats evolve through natural successions, rather than remaining static.

The value of our rivers and coasts is very diverse, ranging from the tourism and recreation importance of features such as sand dunes, beaches and navigable waterways to the scientific and educational benefits of river washlands, coastal cliffs and mudflats. Here, the Government's environmental strategy emphasises the importance of **stewardship** which must underlie the use of environmental resources; balancing the need for economic growth and prosperity with the safeguarding of the natural world (DoE, 1990; Secretary of State for the Environment and others, 1994). In many instances this "**sustainable development**" will involve working with natural processes rather than opposing them through engineering works that sacrifice environmental value for a less hazardous setting. The importance of natural processes is now widely appreciated by coastal defence operating authorities and conservation groups. Indeed, this view is central to English Nature's Campaign for a Living Coast (English Nature, 1992).

Erosion and deposition can also play an important role in minimising the impact of extreme events such as floods, by building up "natural defences". For example, many coastal landforms offer a degree of protection against coastal flooding. **Sand dunes** serve as a natural barrier against high water levels; this has long been an effective coastal defence for many communities around the coast. **Beaches** and **shingle ridges** absorb as much as 90 of the wave energy arriving at the coast by continuously adjusting their form, providing an important component at sea defences where they front embankments or sea walls. **Saltmarshes** and **mudflats** are also effective in dissipating wave energy. All these landforms are dependent upon a continued supply of sediment to maintain their form, often from eroding cliffs; disruption of sediment transport can, therefore, lead to an increase in the degree of risk of erosion behind the sediment–starved landforms elsewhere on the coast.

Erosion, Deposition and Flooding as a Planning Issue

Prior to the mid 1980's, local planning authorities frequently viewed natural hazards such as erosion and flooding as technical problems that the landowner and developer needs to overcome or the responsibility of coast protection authorities and drainage authorities; they were not seen to be land use planning issues. Since the mid 1980's there has been a notable change in perception about the way in which problems are managed. These changes reflect a growing appreciation that the past approach was not in the public interest:

- development in vulnerable locations can lead to demands for expensive publicly funded defence works;

- the possible adverse effects of development on the level of erosion or flood risk elsewhere;

- defence works can have significant adverse effects on the interests of other users of rivers or the coastal zone;

- defence works can encourage **further** development in vulnerable areas, **increasing** the potential for greater losses **when** extreme events occur.

The change in attitude also reflects **concern about** the possible effects of global warming **and sea** level rise and, at a local level, the **effects of recent** major hazard events such as: the **North Wales** floods of February 1990; the Tayside **floods of** 1990 and 1993; and the Holbeck Hall coastal landslide of June 1993.

The Government has advised planning authorities in England and Wales that it is the **purpose of the** planning system to "regulate the development and use of land in the public interest" and **that planners** need to take into account "whether the proposal would unacceptably affect amenities **and the** existing use of land and buildings which **ought to** be protected in the public interest" (PPG1, DoE 1992a; PPG12, DoE 1992b). **Clearly development** in vulnerable areas is not in the **public interest** if appropriate preventative or precautionary measures have not been taken, or if they lead to significant adverse effects on the environment or other interests. The potential impacts of development proposals on the public interest are clearly land use planning issues, especially when public funds are sought later to protect against natural hazards.

In certain areas physical processes, in particular river and coastal flooding **and coastal erosion,** can impinge **directly** on the land use planning functions of local planning authorities. This will include situations where proposals for new development or redevelopment take place in areas which may be at risk, or where proposals to protect existing areas of land from physical processes can have an impact on other planning objectives.

Erosion, deposition and flooding can also generate **indirect** land use issues. Bank erosion, together with sediment delivered to watercourses from hillslopes, can lead to sedimentation problems; the response, maintenance dredging, can cause land use issues when the material has to be disposed of on land or if the operations threaten to damage conservation interests. Conversely, the cessation of dredging following the closure of port and harbour facilities may lead to the degradation of the riverbank environment in city centres (e.g. on the Clyde) or cause an increase in flood risk. These matters can clearly be considered as planning issues and may need to be addressed by local planning authorities.

The planning system clearly has an important role in hazard management, most notably through:

- avoiding unsuitable areas or specifying restrictions on housing occupancy in risk areas;

- ensuring that precautions are taken to prevent runoff from new developments increasing flood risk;

- ensuring that development does not adversely affect floodplain storage and, hence, increase flood risk;

- facilitating the disposal of dredging material from navigable waterways by identifying suitable disposal sites;

- ensuring that development does not affect coastal cliff stability or lead to an increase in coastal erosion;

- ensuring that flood and coastal defence works do not have significant adverse effects on other interests;

- supporting environmental management objectives by ensuring that development does not lead to a decline in value of conservation sites dependant on continued operation of physical processes;

To support decision making in areas affected by these processes, planners will need to have access to reliable background information. However, a planning authority does not owe a duty of care to landowners when granting planning permission and is not liable for losses caused to an adjoining landowner by permitting development. The responsibility and subsequent liability for safe development and secure occupancy rests with the developer or landowner. Any subsequent purchasers should make the necessary enquiries to satisfy themselves about the suitability of the land.

The Need for Adequate Investigation

There are a number of key areas where erosion, deposition and flooding has had a significant impact on land use and development:

- slope erosion and mudfloods on hillslopes;
- dust storms on agricultural land;

- flash floods in upland areas, downstream of reservoirs and dams, and in urban areas with inadequate storm drainage systems;
- bank erosion, sedimentation and channel instability on rivers;
- floods on lowland rivers;
- sedimentation in estuaries;
- floods in low lying coastal areas
- erosion of coastal cliffs;
- wind blown sand in coastal dunes.

These processes should not be considered in isolation as they are often linked, creating complex problems that can affect many different interests. River erosion may oversteepen slopes and lead to instability; erosion of dunes or mudflats can reduce the effectiveness of "natural" coastal defences and may result in more frequent flooding. In turn, the large volumes of fast flowing water in many floods can cause extensive erosion and, when velocities drop, considerable deposition within a channel, leading to navigation problems.

Erosion, deposition and flooding clearly result in significant impacts on society; these problems are often heightened because they impose themselves on unsuspecting or unprepared communities. However, the problems can be readily foreseen, vulnerable areas defined and potential losses significantly reduced by appropriate management responses. It is clear, therefore, that unnecessarily large costs are being incurred by the British economy because of too limited consideration of hazard potential in the development and land use processes.

It is in the developer's interest to determine whether a site is in a vulnerable area, as the possibility of problems will affect the value of the site and its development costs. However, in recent years expenditure on investigations has declined, with many investigations based on minimum costs and maximum speed (ICE, 1991). This has inevitably led to poorly suited designs with inadequate protection, with the risk of damage or losses when major erosion or flood events occur. The solution is not just to spend more. In many cases, greater benefits can arise by better planning of investigations and seeking appropriate specialist advice when necessary.

If there are any reasons for suspecting potential problems at a site a developer should instigate investigations to determine whether:

- the land is capable of supporting the loads to be imposed;

10

- the development will be threatened by erosion, deposition or flooding problems;

- the development will affect the level of risk from erosion, deposition and flooding problems to property, infrastructure and services in adjacent areas or elsewhere within a catchment or coastal system.

Although the majority of questions facing developers relate to site specific issues, the latter point highlights the need to be aware of how the effects of development at a particular location can be transmitted over a wider area and vice versa. Here, an understanding of how processes operate within large physical systems; river catchments and coastal systems is necessary. Indeed, whilst the processes are readily apparent and well appreciated at a local level, they are not often regarded as the product of broader controls and influence the behaviour of river catchments or coastal system. For example, to understand the flood character of a river, something must be known of the climatic, geological, topographic and land use controls on the supply of water and sediments from the surrounding hillslopes. On the coast, the development of beaches or shingle banks need to be seen as the product of sediment transport within dynamic coastal systems. Adequate investigations can also help to ensure that the development is sustainable and that any associated defence works do not adversely affect other environmental interests such as conservation, recreation, tourism and fishing.

Approaches to Investigation

The types and quantities of information needed to support decisions by planners, developers, operating authorities and others with an interest in management will reflect:

- the scale and extent of the issues that need to be addressed;

- the nature of the decision making process involved.

Many decision making processes operate at a variety of levels from the formulation of broad strategic policies at national level by Government Departments (e.g. DoE and MAFF), the development of strategic policies at regional and county level (e.g. the NRA, River Purification Authorities, Coastal Defence Groups, County or Regional Councils), the preparation of detailed policies at the district level (e.g. local planning authorities, coast protection authorities) to the consideration of the suitability of specific sites for development (local planning authorities, developers) or defence works (e.g. the NRA, coast protection authorities etc.). Each level and each group will have its own information requirements.

For planners, operating authorities and developers the information requirement flows from a general awareness of management issues in an area to the need for site specific information. As a result a range of approaches to investigation are relevant at different scales or stages in the decision making process (Figures 2.1 and 2.2).

(i)   **general assessment** of erosion, deposition and flooding issues over broad areas;

(ii) **site reconnaissance** surveys at potential development sites. This can involve either a **site review**, by the planning authority, in advance of the allocation of land when preparing a development plan, or a **preliminary assessment** of shortlisted sites by a developer;

(ii) **detailed assessments** of potential issues at selected sites, as part of site investigation procedures.

The extent of the general investigations will depend on the nature of the issues in an area and the extent to which there are pressures to develop these areas. At the site and detailed levels, the size and nature of the proposed works will influence the scope of the investigations, together with the nature of the site. There will, of course, be overlap between the three levels of investigation; occasionally the stages will be undertaken out of sequence with site reconnaissance, for example, carried out without the benefit of a general assessment. However, reliable detailed investigation depends on good site reconnaissance which in turn benefits from sound general assessment. In this way the detailed **site** information can be analysed and interpreted in the context of an awareness of the broader setting.

Perceptions of Risk

Society is not and cannot be free of risk; it operates within specific levels of tolerance to natural hazards that are defined by both law and common practice. However, the concept of risk and

Figure 2.1 A general model for the investigation of erosion, deposition and flooding issues.

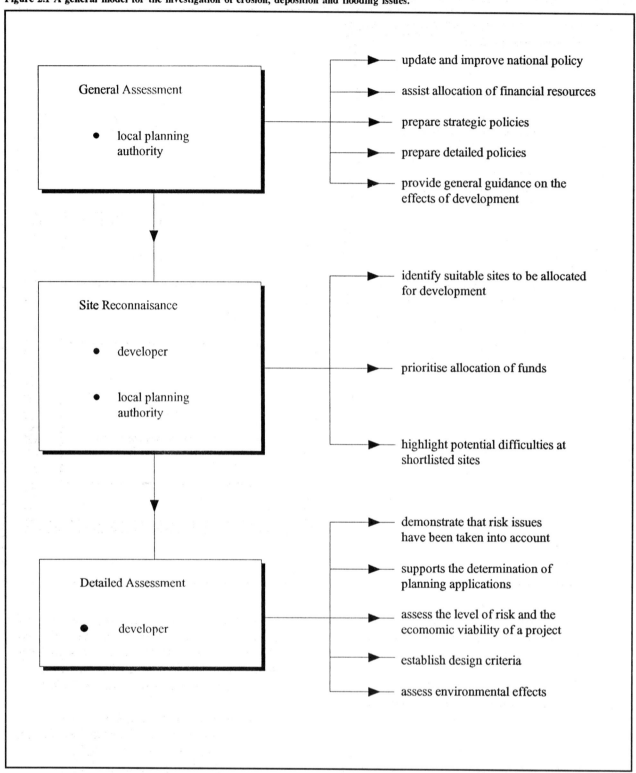

safety tends to be defined vaguely and inconsistently by society, as illustrated by Starr's four "laws" of the acceptability of risk (Starr, 1969):

(i)   acceptability of risk is proportional to the cube of the real or imagined benefits associated with the risk;

(ii)   the public will accept risks derived from **voluntary activities** which are 1000 times greater than those it would tolerate from **involuntary activities** which would generate comparable benefits. Thus, the tolerance of risks created by hazardous sports is thought to be three orders of magnitude greater than that associated with flood events;

Figure 2.2 The sequence of investigations in the planning and development process.

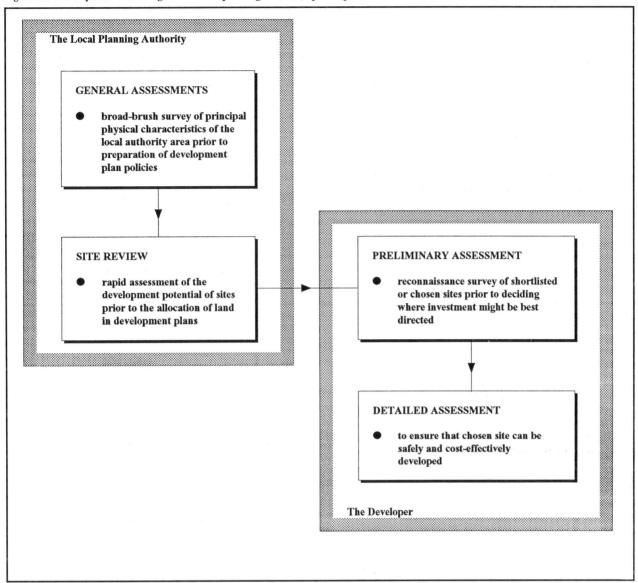

(ii)   the acceptable level of risk is inversely proportional to the numbers of individuals exposed to it;

(iv)   the level of risk tolerated for voluntary accepted hazards is similar to that resulting from disease.

These "laws" highlight the fact that risk involves both objective and subjective elements. **Perceived risk** – the subjective assessments made by individuals – can vary markedly from objective measures of risk. For example, despite the frequent threat of flooding for many parts of the Severn Valley, or repeated coastal landslide movements at Luccombe on the Isle of Wight, people continue to live there. The high level of hazard awareness in these situations, and others, suggests that many residents have made their own judgements that the

potential benefits of living there outweigh the risks; others, of course, may be unable to avoid the risk through financial constraints. The perceived level of risk can be determined by individuals in a variety of ways (Burton and Pushchak 1984):

●   the probability of an event of given magnitude may be calculated by assuming that it is going to be very similar to an event that has happened in the past;

●   perceptions may be anchored to an original experience and not substantially altered by subsequent events.

These judgements, together with objective knowledge, form the basis of risk as it is perceived by "the general public" and define the risk "setting" for particular natural hazards. This "setting" is

Table 2.1 Management responses.

```
Acceptance of the Risk

   •      maintenance and repair
   •      emergency planning
   •      insurance
   •      litigation

Avoiding Vulnerable Areas

   •      land use planning
   •      managed retreat

Reducing the Likelihood of Potentially Damaging Events

   •      regular reservoir inspection
   •      hillslope management
   •      runoff management
   •      floodplain management
   •      river channel management
   •      coastal cliff management
   •      foreshore management

Protecting Against Potentially Damaging Events

   •      early warning
   •      defence schemes
   •      building modifications
```

measured by social scientists in terms of two independent dimensions of risk (Figure 2.3; Slovac et at 1980):

(i)   the degree to which a hazard is **observable**, known to both those exposed and the scientific community, and having an **immediate effect**. Floods and coastal erosion would probably score relatively highly on this dimension;

(ii)   the degree to which the hazard elicits **fear or dread**, the degree to which its consequences are fatal and controllable, and the extent to which individuals are involuntarily exposed. Floods and coastal erosion would have a relatively low score on this dimension as they are frequently controllable with individuals often aware of the potential for risk.

In this context, the risks associated with some erosion and flood hazards can be **voluntary** i.e. willingly accepted by individuals through their own actions. Some floodplain or cliff top homeowners, for example, may elect to buy a home which is often cheaper than the equivalent property in a less vulnerable area; they may also feel there is less need to spend money on house maintenance. Such

decisions are not only a voluntary response to the perceived risk, they are also **economically rational**.

The relationship between voluntary and involuntary risk–taking is important in hazard management as it is central to the notion of **acceptable risk** or **risk tolerance**. Absolute safety is often technically and economically unachievable and, hence, financial and other resources should be allocated to match the threat to society from the hazard. These principles form the basis for MAFF's Indicative Standards of Protection for flood defences in different land use bands see Table 11.7; (MAFF, 1993).

Management of Erosion, Deposition and Flooding Issues

Voluntary acceptance of some level of risk is inherent in every decision to purchase property, carry out economic activity or provide services in vulnerable areas. The resulting costs that can arise when damaging events occur are borne by the affected parties through **maintenance and repair, emergency action** or offset by **insurance** claims or **legal action**. Frequently the acceptance of risk makes sound economic sense. Indeed, in many instances the benefits of placing property, infrastructure and services in vulnerable areas may outweigh the risks from erosion, deposition and flooding. For example, property in vulnerable areas is often cheaper than equivalent property in other areas; here, the savings in property cost can be balanced against the probability of losses or repair costs. Elsewhere, the costs of providing protection measures against, for example, hillslope erosion may considerably outweigh the costs of clear up operations when damaging events occur.

A range of responses are available for managing potential problems associated with erosion, deposition and flooding (Table 2.1):

   •   **Acceptance** of the risk;
   •   **Avoiding** vulnerable areas;
   •   **Reducing** the occurrence of potentially damaging events;
   •   **Protecting** against potentially damaging events.

The range of available responses are outlined in Chapters 8–11. In most cases, however, the response will be complex, involving a variety of measures adopted by residents, landowners and

Figure 2.3 The perceived risk "settings" for erosion, deposition and flooding problems (adapted from Slovac et al, 1980).

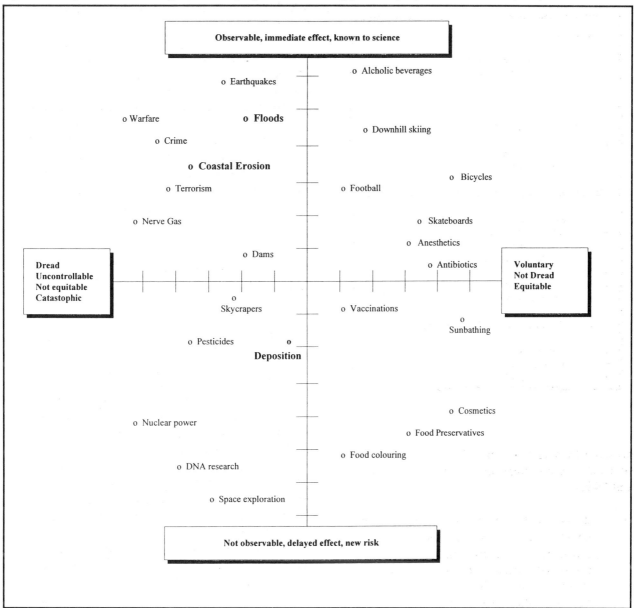

different organisations at different locations within a catchment or coastal system. The scale and complexity of the erosion, deposition and flooding problems dictate that conventional engineering solutions cannot **prevent** the occurrence of potentially damaging events. More realistic aims would involve **reducing** the frequency of damaging events and **minimising** their impact, through:

- providing reliable early warning systems;
- providing appropriate levels of protection to vulnerable communities;
- minimising the effects of human disturbance on the occurrence of erosion, deposition and flood events;
- reducing the damage potential by discouraging inappropriate development in vulnerable areas.

Although erosion, deposition and flooding may present significant **hazards** to land use and development, the processes can also have important roles in the creating or maintaining valuable environmental **resources**. The complexity of the inter–relationships between the processes and management practices, and their combined effects highlight the need for a **coordinated approach** to decision making between land use, flood and coastal defence and conservation interests. This would generally involve **regional** awareness of how the effects of an activity or proposed development at one location can be transmitted over a wide area and create a variety of problems to local amenity or conservation interests, as well as increasing the level of risk to other developments.

The contrast between the broad scale operation of catchments or coastal systems and the limited extent of most local authority areas (Scotland's administrative regions are a notable exception) can lead to management problems, with local authorities concentrating on local issues rather than taking a strategic view of how development can cause problems for their neighbours. Such problems can be overcome through the preparation and use of **management plans**. These plans can provide a mechanism for resolving the conflicts that may arise between flood and coastal defence authorities and conservation agencies. These plans (outlined in Chapter 12) are non-statutory and based on extensive consultation; it is intended that they should present an agreed management strategy that is compatible with maintaining conservation features in a **sustainable condition**.

The most appropriate management option will depend on the nature of the problem, the level of acceptable risk, the availability of resources and the statutory powers available to interested bodies or authorities to tackle the problems. The administrative framework for hazard management is, therefore, a key factor in determining the response to particular problems; this framework is described in Chapter 3.

Chapter 2: References

Acreman M.C. 1985. The effects of afforestation on the flood hydrology of the Upper Ettrick Valley. Scottish Forestry 39, 89–99.

Benson M.A. 1960. Characteristics of frequency curves based on a theoretical 1000 year record. In Flood Frequency Analysis, USGS Water Supply Papers 1543–A.

Best R.H. 1976. The extent and growth of urban land. The Planner 62, 8–11.

Brundtland Commission 1987. Our Common Future. Report of the World Commission on Environment and Development. Oxford University Press.

Burton I. and Puschchak R. 1984. The status and prospects of risk assessment. Geoforum 15, 463–476.

Department of the Environment 1990. This Common Inheritance. Britain's Environmental Strategy. HMSO.

Department of the Environment 1992a. PPG1 General policy and principles. HMSO.

Department of the Environment 1992b. PPG12 Development plans and regional planning guidance. HMSO.

English Nature 1992. Campaign for a Living Coast. English Nature, Peterborough.

Geomorphological Services Limited 1986–87. Review of Landsliding in Great Britain. Reports to DoE.

Inglis C. and Allen F. 1957. The regime of the Thames estuary as affected by currents, salinitiesand river flow. Proceedings of the Institute of Civil Engineers. 7, 827–868.

Institution of Civil Engineers 1991. Inadequate site investigation. Report by the Ground Board, ICE.

MAFF/Welsh Office 1993. Strategy for Flood and Coastal Defence in England and Wales. MAFF Publications.

Rendel Geotechnics, 1995. The Occurrence and Significance of Erosion, Deposition and Flooding in Great Britain. Report to DoE.

Secretary of State for the Environment and others 1994. Sustainable development. The UK Strategy. HMSO.

Slovac P., Fischhoff B. and Lichtenstein S. 1980. Facts and Fears: Understanding Perceived Risk. In R.C. Schwing and W.A. Albers (eds) Societal Risk Assessment: How Safe is Safe Enough? 181–214. New York: Plenum Press.

Starr C. 1969. Social benefit versus technological risk. Science 165, 1232–1238.

Taylor G.G.M. 1970. Ploughing practice in the Forestry Commission. Forest Record No. 73. HMSO.

3 The Legal and Administrative Framework

Introduction

The primary responsibility for dealing with erosion, deposition and flooding issues lies with the landowner, under **common law**. However, a variety of legal and administrative provisions have been introduced through time to contribute to the management of these issues through:

- regulating change and ensuring that activities do not have an adverse effect on the levels of risk to other users;

- improving environmental conditions;

- undertaking mitigation works in the public interest.

In addition to the administrative mechanisms that have been developed to enable property, services and infrastructure to be safeguarded from erosion and flooding (**coast protection; flood defence**), the framework for managing these problems involves aspects of the operation of three broad systems which act to administer and control human activity across the different elements of the landscape and coastal zone (Figure 3.1):

- control of development on **land**;

- regulation of the use of **inland and coastal waters**;

- regulation of activity on the **sea bed**.

Aspects of management also have implications which overlap with administrative mechanisms that have evolved to manage the impact of man on the physical environment:

- protecting the quality of habitats, geological features and landscape (**conservation**);

- maintaining the quality of inland and coastal waters (**pollution control**);

- protecting the water environment from the possible adverse effects of development and land use (**water resource management**).

This Chapter provides a general summary of the framework for management of erosion, deposition and flooding in Great Britain, presenting brief details of individual systems and outlining how they interact to produce a management framework (further details of the framework are presented in the Methodology Report; Rendel Geotechnics, 1995). It should be recognised, however, that distinct legal systems operate for England and Wales, and for Scotland. Separate provisions or statutes are generally prepared for Scotland; these are highlighted in the text where appropriate. However, the management principles outlined in this report are relevant throughout Great Britain, the differences in legal provisions mean that they may be carried out by different authorities and bodies.

Key Elements of the Administrative Framework

British law evolves gradually; the present administrative framework for the management of erosion, deposition and flooding processes must, therefore, be seen to be the product of the way they have presented problems to society in the past. The framework has developed out of a long-standing need to tackle conflicts between an individual's or community's need for protection or maintenance, the restriction of common law rights for the general good and obligations to take into account the interests of other groups e.g. conservation, fisheries and recreation. Since the Earth Summit, held in Rio de Janeiro in 1992, it is

Figure 3.1 The framework for the management of erosion, deposition and flooding issues.

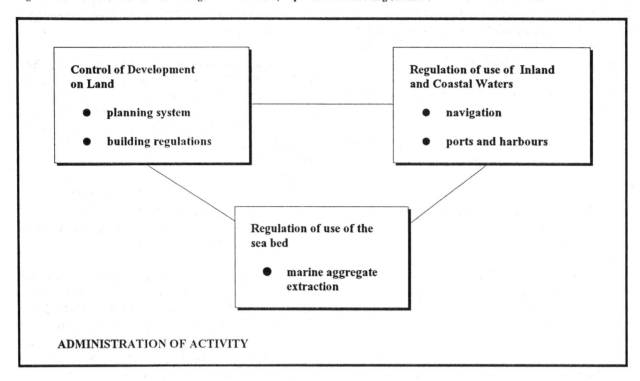

ADMINISTRATION OF ACTIVITY

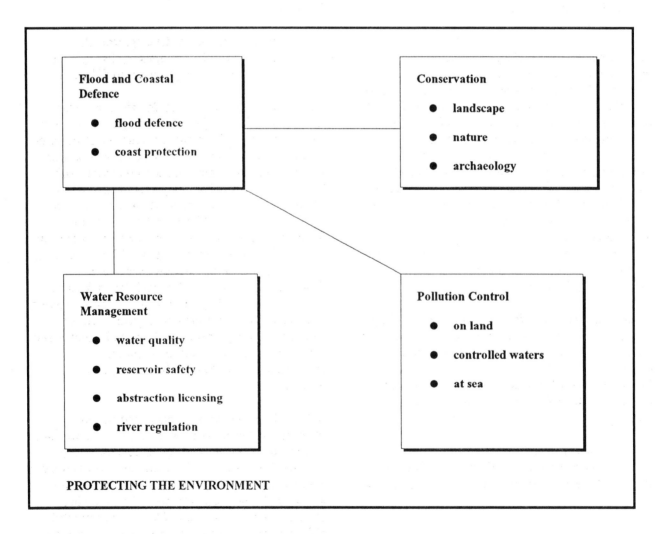

PROTECTING THE ENVIRONMENT

important to see the framework as a mechanism for achieving **sustainable development** with management decisions taken with proper regard to their environmental impact (Secretary of State for the Environment and others, 1994).

The framework is based on a number of underlying principles:

- the primary responsibility for protection or maintenance rests with the owner;

- powers vested with statutory bodies for protection works are permissive and not mandatory; their use is optional rather than obligatory;

- powers vested with bodies such as navigation authorities to maintain waterways are mandatory, i.e. the bodies have a duty to ensure that dredging works are undertaken;

- protection or maintenance works should not have adverse effect on environmental resources.

The complexity of existing arrangements is highlighted by the wide range of bodies and authorities with responsibilities that address or can be influenced by erosion, deposition and flooding problems (Table 3.1). The variety of policies, decisions and actions made by these numerous bodies with an interest in aspects of management can influence the way in which management objectives are realised or achieved. In this context, neither the **flood and coastal defence system** or the **planning system** – the cornerstones of management in Britain (see below) – can be considered to operate in isolation and, hence, their interrelationships with other "systems" is of particular significance.

Flood and coastal defences can be provided by various operating authorities where such works are seen to be in the national interest. The powers contained in the Acts listed below are **permissive** in that they allow works to be carried out, but do not require them (i.e. they are not mandatory powers):

- **Coast Protection Act** 1949;
- **Flood Prevention (Scotland) Act** 1961;
- **Water Resources Act** 1991 (England and Wales);

- **Land Drainage Act** 1991 (England and Wales; as amended by the **Land Drainage Act** 1994).

Schemes are promoted by operating authorities (the NRA, Internal Drainage Boards, local authorities in England and Wales; Regional Councils in Scotland) and may be grant-aided by the relevant Government department (MAFF, the Welsh Office or the Scottish Office). In England and Wales, schemes are evaluated on a combination of technical, economic and environmental factors, as outlined in the recent strategy for Flood and Coastal Defence (MAFF/Welsh Office 1993). There is no provision for compensating landowners where schemes are not undertaken.

The **planning system**, as defined by the **Town and Country Planning Act 1990** (as amended by the **Planning and Compensation Act 1991**) and the **Town and Country Planning (Scotland) Act 1972**, aims to regulate the development and use of land in the public interest. Planning powers are exercised by local planning authorities whose functions include:

- the preparation of development plans;
- the control of development, through the determination of planning applications.

The planning system can be described as "plan-led" in that all planning decisions (either by local authorities or the Secretary of State) must be made in accordance with the development plan, unless material considerations indicate otherwise. This has, in effect, introduced a presumption in favour of those development proposals which conform with the development plan; policy omissions, therefore, can result in development proceeding in areas which might not be suitable. The preparation of development plans, and the exercise of development control, enable decision makers to weigh and reconcile priorities in the public interest. They can ensure that the development needed to help the economy grow, and to provide homes and jobs, takes place in a way that respects environmental constraints and conservation interests.

In both the preparation of plans and handling planning applications, local authorities are required to **consult** with a wide range of interested bodies (e.g. the NRA and conservation agencies) and are advised to take account the Government's policy guidance. In the present context the following guidance is relevant to considering the role of the planning system in managing erosion, deposition

Table 3.1 Bodies with an interest in hazard management.

British Waterways Board
Cadw (Wales)
Conservancy authorities
Countryside Commission
Countryside Council for Wales
Crown Estate Commissioners
Department of the Environment
Department of Transport
English Heritage
English Nature
Harbour Authorities
Historic Scotland
Internal Drainage Boards
Joint Nature Conservation Committee
Landowners
Local Authorities
Ministry of Agriculture, Fisheries and Food
National Rivers Authority
Navigation Authorities
Royal Commission on the Historical Monuments of England (and equivalents in Scotland and Wales)
River Purification Authorities (Scotland)
Scottish National Heritage
Scottish Office Agriculture and Fisheries Department
Scottish Office Environment Department
Water companies and other reservoir owners
Welsh Office

and flooding issues (Table 3.2):

● PPG14 Development on Unstable Land (DoE, 1990);

● PPG20 Coastal Planning (DoE, 1992b);

● Circular 30/92 Development and Flood Risk (DoE, 1992c).

This advice is applicable only in England and Wales; the Scottish Office has not issued advice on these matters (NB: Draft guidance on flood risk was issued by the Scottish Office in March 1995).

In determining planning applications, local planning authorities must take into account representations made in response to statutory consultation or publicity. Permission may be granted subject to such conditions as the local planning authority or the Secretary of State may think fit, provided the conditions are necessary, relevant to planning, relevant to the development, enforceable, precise and reasonable in all other respects. Where matters which are necessary in planning terms cannot be dealt with by way of conditions, such as where the action is not reasonably within the power of the applicant to secure, legal agreements under the 1990 Act may be a necessary precursor to the granting of planning permission.

Applicants may appeal to the respective Secretary of State against a decision to refuse permission or grant it subject to conditions. The Secretary's of State may require applications to be referred to them for decision; this **call-in** power is not frequently exercised and is generally only used where planning issues of more than local importance are involved.

Not all development requires specific planning permission. In England and Wales, the **Town and Country Planning General Development Order 1988** (GDO; as amended) gives general planning

Table 3.2 A summary of the relevant planning guidance in England and Wales.

PLANNING GUIDANCE	CONSIDERATION OF RISKS
PPG14 Development on Unstable Land	"It is important that the stability of the ground is considered at all stages of the planning process. It therefore needs to be given due consideration in development plans as well as in decisions on individual planning applications." (DoE, 1990).
PPG20 Coastal Planning	"Due to the nature of coastal geology and landforms, there are risks, particularly from flooding; erosion by the sea; and land slips and falls of rock. The policy in these areas should be to avoid putting further development at risk. In particular, new development should not generally be permitted in areas which would need expensive engineering works, either to protect developments on land subject to erosion by the sea or to defend land which might be inundated by the sea. There is also the need to consider the possibility of such works causing a transfer of risk to other areas." (DoE 1992b)
Circular 30/92 Development and Flood Risk (Welsh Office Circular 68/92; MAFF Circular FD 1/92)	"Where flood defence considerations arise they should always be taken into account by local planning authorities in preparing development plans and in determining planning applications." (DoE, 1992c).

permission in advance for 28 categories of defined classes of development set out in Schedule 2 to the Order. Such permitted development rights can be withdrawn by a direction under Article 4 of the GDO which normally requires the Secretary of States approval. However, it is intended that these rights should only be withdrawn in exceptional circumstances.

Consenting and Enabling Powers

Management of erosion, deposition and flooding processes is dominated by the need to reconcile a number of conflicting demands:

- protecting vulnerable communities;
- facilitating the provision of competitive ports and navigable waterways;
- meeting the demands of the rapidly-expanding tourism and recreation industries;
- protecting areas of scenic, geological or ecological importance;
- protecting the marine environment.

Finding the right balance may lead to disputes. Some existing communities may feel that conservation interests are creating unnecessary levels of risk to property in vulnerable areas, whilst others may feel that areas of national importance have been progressively spoilt by inappropriate development and the subsequent need for flood or coastal defences. Acceptable solutions to these conflicts can be reached by ensuring that those bodies with powers to undertake works to mitigate against the effects of the processes are required to seek consents from regulating bodies who will take account of a broad range of interests before approving the proposed operation.

In summarising the complex nature of the legislative framework it is useful to separate the provisions into **enabling** and **consenting** elements:

(i)   **Enabling** elements provide the powers for individual authorities to carry out necessary works or maintenance. These include:

- common law rights of landowners to carry out soil conservation, river works and coastal defence works;

- powers for the NRA in England and Wales to carry out flood defence works on main rivers and

the coast, under the **Water Resources Act 1991**;

- powers for IDBs in England and Wales to carry out flood defence works in specified districts with special drainage needs, under the **Land Drainage Act 1991** (as amended by the **Land Drainage Act 1994**);

- powers for local authorities in England and Wales to carry out flood defence works on non-main rivers, under the **Land Drainage Act 1991** (as amended by the 1994 Act) and coast protection, under the **Coast Protection Act 1949**;

- powers for local authorities in Scotland to carry out flood defence works on non-agricultural land, under the **Flood Prevention (Scotland) Act 1961** and coast protection, under the **Coast Protection Act 1949**;

- powers for local authorities to undertake emergency and disaster response measures, under the **Local Government Act 1972 S.138**;

- powers for highways authorities to carry out drainage works (**Highways Act 1980** S.100-101) and the duty to remove soil that obstructs the highway (1980 Act S.150);

- powers for highways authorities in Scotland to protect against erosion and flooding, under the **Roads (Scotland) Act 1984**;

- the duty of the British Waterways Board to maintain commercial and cruising waterways in a suitable condition, under the **Transport Act 1968** (S.105);

- the duty on owners or operators of reservoirs with a capacity of over 25,000m^3 above the level of adjacent natural ground, to employ qualified engineers to maintain safety standards, under the **Reservoirs Act 1975**. Local authorities have a duty to ensure that the provisions of the Act are followed;

- powers for harbour authorities to undertake a range of flood and erosion control works, dredging and channel improvements, under local and private Acts or by means of harbour orders under the **Harbours Act 1964** (S.14, S.16).

(ii) **Consenting** elements ensure that proposed works or activities take appropriate account of other interests. These include:

- Land drainage consents under the **Water Resources Act 1991** or **Land Drainage Act 1991**, or NRA and IDB byelaws made under these Acts;

- approval of the relevant coast protection authority under the **Coast Protection Act 1949** (S.16);

- consent from the relevant coast protection authority for the removal of materials from the sea bed or foreshore in areas where a **Coast Protection Order** has been made under the 1949 Act S.18;

- licences issued by MAFF under the **Food and Environment Protection Act 1985** Part II for waste disposal and construction works below HWM;

- Waste management licences issued by waste regulation authorities for disposal on land, under the **Environmental Protection Act 1990**;

- Consents for discharges into controlled waters issued by the NRA under the **Water Resources Act 1991**;

- the granting of specific planning permission for development, made by local planning authorities under the **Town and Country Planning Act 1990 (1972 Act in Scotland)** or general permission granted by

the General Development Order 1988;

- the approvals issued by the local authority under the **Building Regulations** (this procedure is intended to ensure the safety of a building and does not consider broader implications of a proposed design);

- the **Government View Procedure** for examining proposals for marine aggregate extraction, coordinated by the Department of the Environment, the Welsh Office and Scottish Office;

- Notification of **Potentially Damaging Operations** likely to damage SSSIs by the relevant conservation agency, under the **Wildlife and Countryside Act 1981**;

- Consent for works on or around scheduled ancient monuments, made by the respective Secretary of State under the **Ancient Monuments and Archaeological Areas Act 1979**;

- consent for depositing materials in a restricted area around a vessel of historical or archaeological value, under the **Protection of Wrecks Act 1973**.

These consenting elements are supported by the provision of **environmental duties** to many of the bodies with powers to undertake works in the river or coastal environment (Table 3.3). In general, these duties require the operating bodies to have regard for environmental interests; the NRA and IDB's however, have a duty to further the conservation and enhancement of natural beauty, habitats and geological features in so far as this is consistent with their functions.

An important additional element within the framework are the **formal linkages** between authorities, bodies, managers and users through which decision makers can take into account the views of other interests. These linkages provide the means for:

- anticipating problems before they arise;

Table 3.3 Environmental duties of authorities other than nature conservancy bodies.

Authority	Environmental Duty
NRA and Internal Drainage Boards	– to further the conservation and enhancement of natural beauty and the conservation of flora, fauna and geological or physiographical features of special interest so far as this is consistent with their functions. – to have regard to the desirability of protecting and conserving buildings, sites and objects of archaeological, architectural or historic interest. – to take into account any effect which the proposals would have on the beauty or amenity of any rural or urban area or on any such flora, fauna, features, buildings, sites or objects (Land Drainage Act 1991 S.12). The NRA's environmental duties are defined by the Water Resources Act 1991 S.2 (duty to promote conservation) and S16(1) (duty to further conservation).
Local authorities	– have a duty to have regard to the environment and conservation when carrying out land drainage and flood defence work (Land Drainage Act 1994)
Fisheries Department (with regard to determining dumping at sea licences)	Licensing authorities must have regard to the need to: – protect the marine environment, the living resources which it supports and human health. (Food and Environment Protection Act 1985 S.8).
Harbour Authorities	Harbour authorities must have regard to: – the conservation of natural beauty of the countryside, flora, fauna and geological and physiographical features of special interest. – preserving freedom of access to places of natural beauty. – maintaining the availability to the public to visit or inspect buildings, sites or objects of interest. (Harbours Act 1964 S.48A a S.48A, as amended in 1992).

- coordinating management decisions to ensure that users do not interfere with other interests;

- ensuring that management reflects local needs as well as national policy;

- resolving conflicts when they arise.

The Physical Basis for Planning and Management

In recent years it has been widely appreciated that the contrast between the extent of administrative units and the broad scale or operation of erosion, deposition and flooding processes has been a significant constraint to effective planning and management. Flood and coastal defence issues, for example, are now addressed within the context of catchments or coastal systems, with **management plans** currently being prepared by various organisations or groups to establish broad strategies within which local management decisions can be made (Table 3.4).

The Framework for Catchment Management

The NRA, as guardians of the water environment, have a major role in catchment management in England and Wales. As stated in their 1990/91 Corporate Plan, they are committed to a programme of preparing and implementing **catchment management plans** to integrate the range of their water resource functions such as water quality, flood defence, fisheries, recreation, conservation and navigation. A national programme is currently in place to prepare 164 plans by 1998. The plans consider the various water users' interests and develop a long term "vision" and medium–term strategies and actions through consultations with local communities and organisations, highlighting key issues and developing practical solutions. Pickles and Woolhouse (1994) note that, to date, local authorities are finding the plans

Table 3.4 Summary of the purpose and nature of management plans.

	Shoreline Management Plan	Coastal Zone Management Plan	Catchment Management Plan	Water Level Management Plan
Primary Purpose	Strategic planning for coastal defence.	Framework to help resolve competing pressures in the coastal zone.	Strategic planning of NRA functions.	Establish an agreed balance between agriculture, flood defence and conservation in wetland areas.
Main Issues Covered	Coastal Defence	Development Recreation Landscape Environment Navigation Coastal defence	Flood defence Water resources Navigation Conservation Fisheries Pollution	Flood defence Conservation Agriculture
Extent and Boundaries	Sediment cell/sub–cell	Varies: May focus on specific areas such as estuaries or longer stretched of open coast.	Main river catchments.	Wetland areas, especially SSSIs.
Lead Authority	Maritime District Council or NRA (in collaboration with other members of the relevant coastal group).	District or County Council, Borough Council or other interested organisation.	NRA	NRA

particularly useful for establishing joint initiatives with the NRA.

Water level management plans are prepared by drainage authorities in England and Wales to establish an agreed balance between agriculture, flood defence and conservation in wetland areas. Conservation guidelines issued by MAFF state that plans should be produced for areas where water levels are managed, with priority given to SSSIs (MAFF/DoE/Welsh Office, 1991; see also MAFF et al, 1994).

There are no formal catchment management arrangements existing in Scotland. This is largely a result of the fragmented approach to water resource management throughout Scottish river basin systems, the reliance upon actions by riparian owners and the absence of funds for the purpose. Perhaps most significantly, the Flood Prevention (Scotland) Act 1961 restricts the powers available to a Regional Drainage Authority strictly to urban areas, and the Land Drainage (Scotland) Act 1930 enables agricultural drainage measures and watercourse maintenance and the like to be carried out by landowners. An informal approach to catchment management has been set up by the Tweed River Purification Board – The Tweed Forum – which has attracted wide voluntary

interest from statutory bodies, commercial organisations, recreational groups and landowners. The objectives of the Forum, acting as an informal liaison group, are to promote a unified approach to the individual members activities in the spirit of catchment management but without formal guidelines or external funding. The Forth Estuary Forum has recently been established, initiated by the RSPB primarily as a conservation forum; the group does not attempt to address overall river management objectives nor does it embrace the entire catchment area.

The Framework for Coastal Management

Coastal management has been the focus of considerable interest in recent years, reflecting concerns about balancing social and economic growth, the need for coastal defences and protection of the natural environment (see, for example, House of Commons Environment Committee, 1992; DoE, 1992a; Rendel Geotechnics, 1993). In response the Government has promoted and encouraged the preparation of a variety of management plans in England and Wales addressing different central issues and involving different key organisations:

Table 3.5 The coastal groups in England and Wales (as of October 1994).

```
-    West Cumbria Coastal Group
-    River Ribble to Morecambe Bay Coastal Group
-    Llandudno to Mersey Estuary Coastal Group (The Liverpool Bay Group)
-    Tidal Dee User Group
-    Ynys Enlli to Llandudno Coastal Group
-    Cardigan Bay Group
-    Carmarthern Bay Coastal Engineering Study Group
-    Swansea Bay Coastal Group
-    Severn Estuary Coastal Group
-    Somerset and Avon Coastal Group
-    Devon Coast Protection Advisory Group
-    Cornwall Countywide Coast Protection Group
-    Standing Conference on Problems Associated with the Coastline (SCOPAC)
-    East Sussex Coastal Liaison Group
-    Kent Coastal Group
-    Anglian Coastal Authorities Group
-    Humber Estuary Coastal Authority Group
-    North East Coastal Authorities Group
-    Northumbrian Coastal Group
```

A MAFF/Welsh Office Coastal Defence Forum was established in 1991, involving representatives of the coastal groups and spokesmen from the NRA and English Nature. The Forum's terms of reference are:

"To provide a national forum on coastal defence, including sea defence and coast protection matters, for government officials, representatives of Coastal Groups and others with an interest in England and Wales."

The Forum aims to assist authorities in their coastal defence functions by:

```
-    further cooperation between parties with responsibility for coastal defence;
-    sharing experience, data etc.;
-    identifying best practice;
-    identifying research needs and possibilities;
-    promoting strategic planning and coastal defence management;
-    identifying obstacles to progressing planned works;
-    keeping abreast of policy developments, R&D results and new initiatives.
```

(i)   **Coastal Management Plans and Estuary Management Plans**; aimed at encouraging the management of all aspects of the human use of the coast to yield the greatest benefit to the present population whilst maintaining the potential of coastal systems to meet the needs and aspirations of future generations (DoE, 1993). Such plans are viewed by the Government as best directed at resource management issues (e.g. recreation, conservation, ports, fisheries, land use, water quality, tourism etc.); whilst the avoidance of erosion or flood risk areas may be considered, the plans are not seen as the place for defining a coastal defence strategy, although the plans should take account of existing shoreline management strategies. Consequently the key interest groups would be:

●   local planning authorities
●   MAFF
●   the NRA
●   conservation agencies and groups
●   harbour authorities
●   sea fisheries committees
●   recreational groups and bodies
●   industrial users

(ii)   **Shoreline Management Plans**; MAFF and the Welsh Office have encouraged the setting up and operation of **coastal defence groups** comprising coast protection authorities, drainage authorities and other bodies with coastal responsibilities (MAFF 1993; Table 3.5; Figure 3.2). These groups are encouraged to prepare shoreline management plans which take into account natural processes, planning pressures, current and future land use, defence needs and environmental considerations on specific lengths of coast, preferably littoral cells or sub-cells (Figure 3.3; MAFF 1994). These plans are intended to integrate

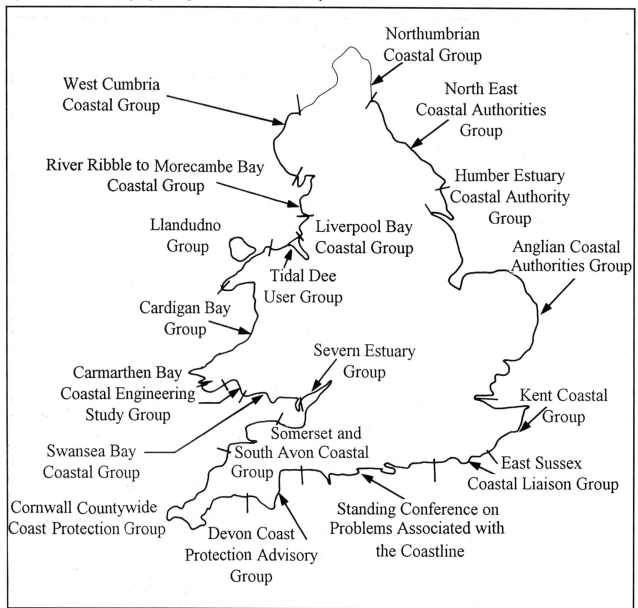

coastal defence and planning issues, to ensure that statutory development plans can take account of coastal erosion and flooding problems, and to avoid inappropriate development. The key interest groups are, therefore:

●   MAFF
●   coast protection authorities
●   the NRA
●   local planning authorities
●   conservation agencies

Chapter 3: References

Department of the Environment 1990. PPG14 Development on Unstable Land. HMSO.

Department of the Environment 1992a. The Government's Response to the Second Report from the House of Commons Select Committee on the Environment – Coastal Zone Protection and Planning. HMSO.

Department of the Environment 1992b. PPG20 Coastal Planning. HMSO.

Department of the Environment 1992c. Circular 30/92 Development and Flood Risk (MAFF Circular FD 1/92; Welsh Office Circular 68/92). HMSO.

Department of the Environment 1993. Managing the coast: a review of the coastal management plans in England and Wales and the powers supporting them. DoE.

House of Commons Environment Committee 1992. Coastal Zone Protection and Planning. HMSO.

HR Wallingford 1993. Coastal Management: Mapping of littoral cells. Report SR 328.

MAFF/DoE/Welsh Office, 1991. Conservation guidelines for drainage authorities.

MAFF 1993. A Strategy for Flood and Coastal Defence in England and Wales. MAFF Publications.

MAFF 1994. Shoreline management plans: a guide for operating authorities. MAFF Publications.

MAFF, Welsh Office, Association of Drainage Authorities, English Nature and NRA 1994. Water level management plans: a procedural guide for operating authorities. MAFF Publications.

Pickles L. and Woolhouse C. 1994. Catchment management plans: a strategic view of flood defence. In Proc. MAFF Conference of River and Coastal Engineers.

Rendel Geotechnics 1993. Coastal Planning and Management: a review. HMSO.

Rendel Geotechnics 1995. Erosion, Deposition and Flooding in Great Britain. Methodology Report. Open File Report held at the DoE.

Scottish Office Environment Department 1995. Planning and Flooding. National Planning Policy Guidance (draft).

Secretary of State for the Environment and others, 1994. Sustainable development. The UK Strategy. HMSO.

4 Information needs for Sustainable Planning and Management

Introduction

The UK strategy for achieving sustainable development (Secretary of State for the Environment and others, 1994) is based on a number of specific principles:

- decisions should be based on the best possible scientific information and analysis of risks;

- where there is uncertainty and potentially serious risks exist, precautionary action may be necessary;

- ecological impacts must be considered, particularly where resources are non-renewable or effects may be irreversible;

- cost implications should be brought directly to the people responsible – the "polluter pays" principle.

All of these principles are relevant to the management of the issues generated by erosion, deposition and flooding processes (see Chapter 2). Indeed, the broad scale operation of these processes dictate that catchment and coastal system – wide management is necessary to ensure that land use and development is sustainable. In this context management requirements can be addressed in terms of three **key considerations**:

- the **risks** associated with geomorphological processes, as highlighted in Chapter 2 (see also Rendel Geotechnics, 1995);

- the **sediment budget** of a catchment or coastal system which can be a major factor in ensuring the sustainability of natural and engineered defences, navigation uses, port and harbour operations, conservation sites

and recreation areas, and operations such as aggregate extraction;

- the **sensitivity** of the physical environment to natural or man-induced change. This can be an important factor in assessing the potential effects of development, for example, along a river or coastline.

These key considerations are relevant at all stages of the decision making process, from regional or strategic level to the determination of site specific proposals. In Chapter 2 an hierarchical model was set out which indicated the investigation approach suited to different stages of the decision-making process; from general assessment to site reconnaissance and detailed assessment. As the information required for these investigations changes from a general awareness of the natural environment to the need for site specific information, the questions that should be considered remain broadly the same (Figure 4.1). The significance of some questions and the extent to which they will need to be addressed will, however, vary according to the nature of the environment and the extent to which it is subject to pressure for development.

Before examining specific information requirements in different environments, it is necessary to consider what types of earth science information are needed to address the key considerations and the ways in which such information can be integrated, analysed and presented to answer the specific questions posed by planners and managers. Given the wide variety of environments and associated issues it is inappropriate to be dogmatic about the precise methods that can be employed in data collection exercises. However, in most cases an understanding of **landforms** will be a key component of most investigations. This is because landforms are capable of study at a wide range of scales from individual sites to major catchments and coastal systems, and can often be a sensitive

Figure 4.1 The elements of earth science investigations.

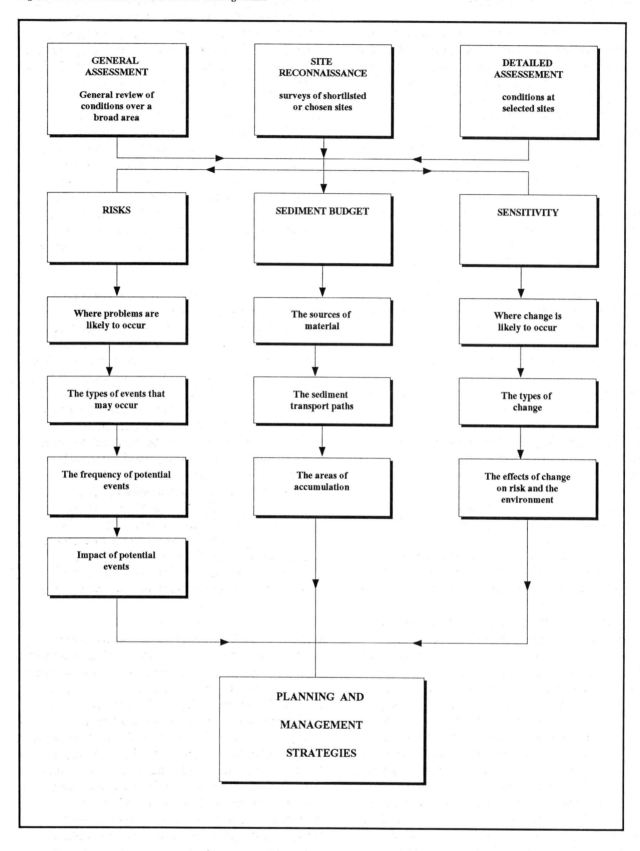

indicator of **materials** and **process.**

The identification and mapping of landforms can provide a spatial framework for understanding site specific information on materials or processes and for extrapolating this information into less well investigated areas. Recognition of particular landforms can also be fundamental to the assessment and prediction of potential environmental change. Indeed, as decision-makers often require immediate knowledge of potential changes without giving opportunity for long-term monitoring, it is frequently necessary to estimate change and processes through the interpretation or modelling of landforms and materials.

However without exception, an **inventory** of landforms alone will not provide the right documents to pass onto planners and managers. This information will need to be **interpreted** in terms of risks, sediment budget or sensitivity, to show features specifically relevant to the management concerns in that area. Thus, for example, management of urban landslide problems may require an assessment of the landslide hazard and the potential consequences of ground movement in map form. Other data, if shown on the maps, may only serve to confuse or divert attention away from the main issues facing planners and managers.

In many instances, it may be difficult for non-technical users to make judgments about the relative significance of information shown on different interpretive maps (ie. risks, sediment budget, sensitivity). Consequently, a series of these derived maps will often need to be integrated and **summarised** to indicate, in general terms, the principal constraints or resources in an area. The end product, essentially a form of **sieve map**, should be familiar and accessible by planners and managers, ensuring that maximum value is gained from the earth science information. Such maps also provide a mechanism for ensuring that the key environmental parameters are expressed in terms of their significance for planning and development. For example, **planning guidance** can be prepared to highlight areas that are suitable for development, have significant constraints that could be overcome by precautionary works or are largely unsuitable on the basis of the risks or the sensitive nature of the environment. Where there are concerns about the possible environmental effects of a proposed development, the collected and interpreted information may need to be summarised by the developer in terms of an **environmental assessment.**

The following sections are intended to outline the general principles of the investigation procedure from **inventory** to **interpretation** and **summary** (Figure 4.2). The procedures are, however, relevant at each of the 3 levels of investigation identified in Chapter 2. The approach will also have wider applicability to other bodies with an interest in aspects of catchment and coastal management, including coastal defence operating authorities and groups, and conservation agencies. Some elements of the approach will, however, become more important for particular end-users. For example, planners are generally best served by summary maps, whereas the technical officers within, for example, a coast protection authority would gain greater benefit from the interpretative or inventory maps.

Before proceeding further it is important to stress that lack of awareness of the potential for damaging events in an area is often a significant obstacle to be overcome, especially where the threat is associated with extremely rare events (e.g. flash flooding in dry valleys and tributary streams) of failure of man-made structures (e.g. dams and embankments). It is important, therefore, that the first step in any investigation should be to recognise that problems can be expected to occur in most administrative areas; the characteristic vulnerable settings shown in Table 4.1 with reference to particular terrain units can help authorities to identify what problems can be expected in particular landscapes (Figure 4.3). Once the range of problems in an area have been anticipated the identification of the extent of vulnerable areas can be achieved from readily available sources (Table 4.2) and specialist bodies (Table 4.3).

It should be noted, however, that on the coast much of this information is currently being collated as part of shoreline management plans which take a strategic view of coastal defence issues along a section of coastline, (Table 4.4). For inland areas, a similar potential source may be catchment management plans. It is envisaged that these plans and the detailed information held by the various operating authorities will often be a major source of earth science information for planners. Many operating authorities have well established monitoring programmes to keep under review changes in the physical character of a coastline; these programmes can also be a source of useful information for planning authorities as they monitor the effectiveness of their development plan policies, avoiding the need for duplication of effort.

Figure 4.2 The investigation procedure.

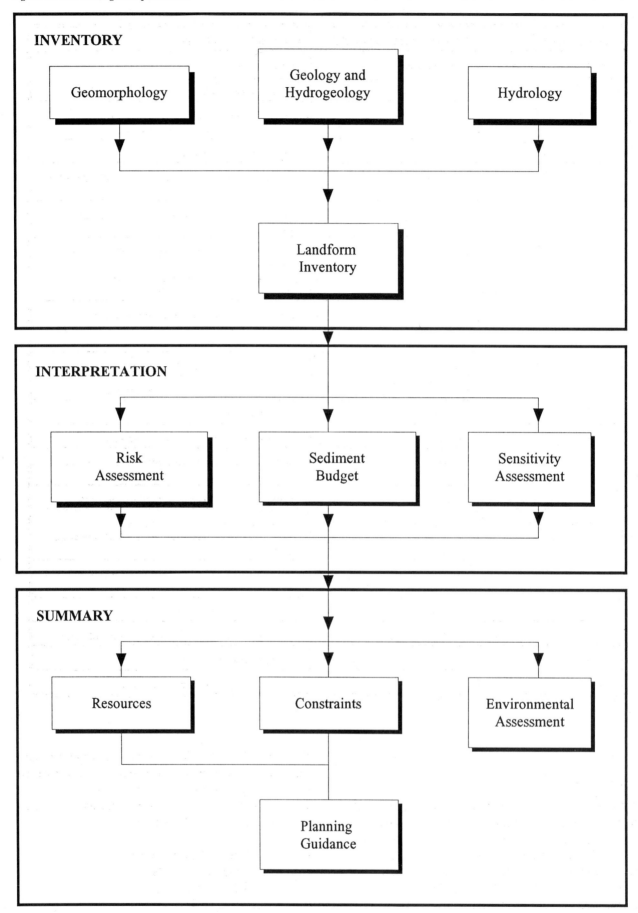

Table 4.1 Characteristic vulnerable settings associated with different terrain units.

TERRAIN UNIT	CHARACTERISTIC PROCESSES	VULNERABLE SETTINGS
Mountainous and Upland areas	Upland soil erosion	Upland peaty soils. Areas of recreation pressure. Recently afforested land.
	Flash floods	Land adjacent to streams. Areas downstream of dams and reservoirs.
	Sedimentation	Upland lakes and reservoirs.
Undulating lowlands	Water erosion	Areas of poor land management associated with silty and fine sandy soils.
	Wind erosion	Areas of poor land management associated with fine sandy soils.
	Flash floods	Land adjacent to tributary streams, especially around the margins of upland areas. Areas with inadequate storm drainage systems. Areas downstream of dams and reservoirs.
	Unstable river channels	Alluvial rivers around the margins of upland areas. Alluvial rivers upstream and downstream of channelisation, river engineering works, bridges etc.
	Mudfloods	Dry valleys in areas of poor land management.
	Lowland floods	Low lying land adjacent to rivers and streams.
	Sedimentation	Within alluvial river channels.
Estuaries	Flooding	Low lying land adjacent to rivers and streams.
	Sedimentation	Estuaries with a flood tide dominance which results in a net movement of sediment into the estuary.
	Channel erosion	Estuary margins developed in soft cohesive materials.
Coastal lowlands	Flooding	Low lying land adjacent to rivers and streams. Low lying coastal land, including areas behind sand dunes, shingle ridges etc.
	Wind erosion	Areas of poor land management associated with lowland peat soils.
	Sedimentation	Within alluvial river channels.
Coastal cliffs	Landsliding and cliff recession	Unprotected soft rock cliffs, especially on exposed coasts.
Sand dunes	Wind erosion	Unstabilised, bare sand areas.
	Flooding	Low lying coastal land behind sand dunes.

Inventories

A first step in most data collection exercises will be to produce an inventory of catchment and coastal features and to characterise these features in terms of materials, form and processes. The best systematic and nation–wide survey of **coastal landforms** was produced by the then Nature Conservancy Council as part of its **Atlas of Coastal Sites Sensitive to Oil Pollution**. However the maps, at 1:100,000 scale, provide only a broad indication of coastal features; there is no equivalent source of information for inland landforms. For many studies, therefore, it will be necessary to seek more detailed information from:

- Ordnance Survey maps;
- Hydrographic charts;
- Aerial photographs;
- Field inspection;
- Consultation with key interest groups.

33

Figure 4.3 Characteristic erosion, deposition and flooding problems in different terrain units.

Terrain Unit	Upland erosion	Water erosion	Wind erosion	River channel instability	Mud floods	Flash floods	Lowland floods	Coastal floods	Sedimentation	Landsliding and cliff recession	Wind blown sand
Mountainous and Upland Areas	•	•		•		•			•		
Undulating Lowlands		•	•	•	•	•	•		•		
Estuaries							•	•	•		
Coastal Lowlands			•					•	•		
Coastal Cliffs										•	
Sand Dunes											•

Table 4.2 Principal sources of general information.

PROBLEM	ENGLAND AND WALES	SCOTLAND
Hillslope erosion	Soil Survey and Land Research Centre: – maps of erosion risk on agricultural land (1:250,000 scale) – maps of soil associations prone to water erosion can be <u>derived</u> from National Soil Maps (1:250,000 scale)	Macaulay Land Use Research Institute: – no published information on erosion prone soils
Wind erosion	Soil Survey and Land Research Centre: – maps of soil associations prone to wind erosion can be <u>derived</u> from National Soil Maps (1:250,000 scale)	Macaulay Land Use Research Institute: – no published information on erosion prone soils
Flash floods	National Rivers Authority: – areas prone to flooding shown on S.105 survey maps (various scales)	Various possible sources, including: – Island and Regional Council Drainage Department – Island and Regional Council Roads Department – River Purification Boards
	Water companies and other owners of major reservoirs generally have flood inundation maps for use in event of dam failure.	
Channel Instability	No systematic collection of records. Channel positions on present day topographical maps can be compared with positions on historical maps.	
Lowland flooding	National Rivers Authority: – areas prone to flooding shown on S.105 survey maps (various scales)	Various possible sources, including: – Island and Regional Council Drainage Department – Island and Regional Council Roads Department – River Purification Boards Floodplains extent can be <u>derived</u> from Macaulay Land Use Research Institute soil maps (1:250,000 scale) or British Geological Survey geological maps (1:50,000/1:63,360 scale).
River and estuary sedimentation	No systematic collection of records. Channel depth on present day Admiralty Charts can be compared with positions on historical charts.	
Coastal flooding	National Rivers Authority: – areas prone to flooding shown on S.105 survey maps (various scales)	Various possible sources, including: – Island and Regional Council Drainage Department – Island and Regional Council Roads Departments – River Purification Boards Flood prone areas can be <u>derived</u> from Macaulay Land Use Research Institute soil map (1:250,000 scale) or British Geological Survey geological maps (1:50,000/1:63,360 scale). Ordnance Survey maps can be used to identify the extent of land below 5m.
Coastal cliff erosion	Department of Environment/Welsh Office/Scottish Office. Distribution of recorded landslides can be obtained from 1:250,000 scale county/Scottish region landslide maps or the National Landslides Databank.	
Mobile sand dunes	Soil Survey and Land Research Centre: – maps of soil associations prone to wind erosion can be <u>derived</u> from National Soil Maps (1:250,000 scale).	Macaulay Land Use Research Institute: – no published information on erosion prone soils Scottish National Heritage: – The Beaches of Scotland; provides general information on the character of beaches, dunes, machairs and links.

Table 4.3 Sources of specialist advice on various erosion, deposition and flooding issues.

SYSTEM	SPECIALIST BODIES AND AUTHORITIES
Hillslopes	Soil Erosion: – Soil Survey and Land Research Centre, Silsoe – Macaulay Land Use Research Institute, Aberdeen – ADAS/SOAFD Flash Flooding: – NRA – River Purification Boards – Institute of Hydrology, Wallingford
Rivers	Channel Instability and Sedimentation: – NRA Lowland Flooding: – NRA – River Purification Boards – Scottish Regional Councils – Institute of Hydrology, Wallingford Flooding from dam failure – BRE Garston, Watford Estuary Sedimentation: – Hydraulics Research, Wallingford – Institute of Estuarine and Coastal Studies, University of Hull
Coast	Coastal Flooding: – NRA – Scottish Regional Councils – Proudman Oceanographic Laboratories Coastal Cliff Erosion: – Coastal defence groups – Coast protection authorities – Rendel Geotechnics (National Landslide Databank) – British Geological Survey – Coast Protection Survey of England (MAFF) Wind Erosion in Dunes – Soil Survey and Land Research Centre, Silsoe – Macaulay Land Use Research Centre, Aberdeen – Joint Nature Conservation Committee (Sand Dune Vegetation Survey)

Individual landforms or groups of landforms should then be **characterised** in terms of their materials and processes, with reference to available sources of information (Table 4.2) or, if necessary, specially commissioned studies. Amongst the most important data sources will generally be:

- Geological maps and memoirs (BGS, 1994);
- Soil maps and memoirs;
- National Reviews of Ground–Related Problems (see Chapter 1);
- The Macro–review of the Coastline of England and Wales (HR Wallingford, various dates);
- The Beaches of Scotland (Ritchie and Mather, 1985);
- Catchment Management Plans;
- Shoreline Management Plans;
- Historical maps and records.

The identification and characterisation of landforms provides an essential basis for understanding the nature of a particular catchment or coastline. The information collected is often best presented as a **Geomorphological Map** portraying some or all of the following;

(i)   **Surface Form;** morphology and slope angle;

Table 4.4 The nature of shoreline management plans in England and Wales (after MAFF, 1994).

Shoreline Management Plans are intended to provide a strategic framework for the management of coastal defences for a specified length of coast. Such plans should take account of natural coastal processes and human and other environmental influences and needs. The aim of a plan is to provide the framework for the development of **sustainable coastal defence policies** within a sediment cell or sub-cell.

The specific objectives of a Plan are:

- to develop an understanding of the coastal processes operating within the sediment cell or sub-cell and their influence on the shoreline;

- to predict the likely future evolution of the coast;

- to identify all the assets within the area covered by the Plan which are likely to be affected by coastal change, including the developed and natural environment, amenity, leisure facilities and other infrastructure;

- to identify means of maintaining and enhancing the natural coastal environment;

- to facilitate consultation between those bodies with an interest in the shoreline;

- to set objectives for the future management of the shoreline;

- to inform the statutory planning process;

- to assess a range of coastal defence policy options;

- to identify the need for regional or site specific research and investigation; and

- to outline future monitoring requirements.

Five key issues are addressed in the preparation of a Plan. These are:

- coastal processes;

- natural environment;

- the human and built environment;

- development in the coastal zone; and

- coastal defences.

(ii) **Geological Conditions;** bedrock lithology, superficial deposits, structural form;

(iii) **Processes;** surface form or materials can be mapped to indicate the dominant process. For example:

- **erosion** can be portrayed by the presence of landslides or eroding cliffs;
- **flooding** can be mapped by identifying the extent of land affected by past floods.

The map can also indicate the nature of **sediment transport processes** by highlighting sediment sources and sinks, together with dominant directions of sediment movement or littoral drift.

A geomorphological map will seldom be an appropriate end product to pass on to planners and managers. Such maps will often appear too complex and many of its features may be meaningless to the non-specialist. The map may also carry data which are not relevant to his immediate concerns and the significance of the various map units to planning and development may be unclear. For many areas, decision-makers will be better served by specifically prepared maps, based on an interpretation of the inventory information to address risk, sediment budget or sensitivity issues.

Risk Assessment

The terminology used when discussing natural events and their impact can be misleading. It is not uncommon for terms like hazard, risk and vulnerability to be used interchangeably by different people. In fact, all three have distinct meanings which can be defined as follows:

- **hazard** describes the chance of a potentially damaging event occurring within an area;
- **risk** describes the possible losses arising as a result of a damaging event;
- **vulnerability** describes the degree of loss or damage to particular sectors of a community (e.g. buildings, infrastructure, services etc.) at risk from an event i.e. different elements will face different levels of risk depending on their vulnerability.

In most instances a risk assessment will involve some or all of the following steps:

(i) **defining hazard zones** which could be affected by erosion, landsliding, flooding etc. This can be achieved by mapping surface form or topography to delineate the extent of vulnerable land;

(ii) **defining hazard potential** in terms of the nature and size of events that could be expected in an area. This will require an appreciation of:

- the nature and magnitude of historical events;
- the theoretical occurrence of events from rainfall or wave climate records;
- causes and mechanisms of possible events;
- factors influencing the pattern of events;
- the effects of development on the incidence of events;
- the standard of protection provided by existing flood and coastal defences.

(iii) **identifying land use elements at risk;** land use within hazard zones can be classified according to:

- the **relative importance** (Table 4.5);

- the **vulnerability of the buildings** expressed in terms of structural vulnerability and material vulnerability (eg. Table 4.6).

(iv) **assessing the potential consequences;** the severity of the potential consequences of each type of event can be considered in terms of a simple 3-fold classification:

- **total loss;** ie. deaths, severe casualties or destruction of property;
- **partial loss;** ie. injury, minor casualties or serious and severe property damage;
- **minor loss;** ie. inconvenience or slight and moderate property damage.

(v) **assessing the likelihood of damaging events;** an appreciation of the size and frequency of potentially damaging events is a central component of risk assessment. Two contrasting approaches are generally used to predict the occurrence of events:

- **deterministic** based on a general understanding of the physical processed operating within a coastal system. For example, erosion of coastal cliffs is generally modelled through **stability analyses** which compare the destabilising and resisting forces acting within a slope to establish a factor of safety (e.g. Bromhead. 1986). The Flood Studies Report (NERC, 1975) presented a number of deterministic methods of estimating flood discharge from specified rainfall and catchment conditions. However, all deterministic approaches suffer, to a greater to lesser degree, from the need to simplify the complexity of the physical environment and make general assumptions.

- **probabilistic** regarding individual events as a random part of a natural series of events of varying magnitude and frequency whose distribution can be established from a sequence of records and whose probability of occurrence in a given period can be calculated using

Table 4.5 Typical values of importance of land use and importance of structures and services (after Cole et al, 1993).

LAND USE CATEGORY		VALUE
1.	Public open space, farmland, tidal land.	0.3
2.	Domestic houses (single family occupancy), secondary communications network/roads and railways, small factories and small places of assembly.	1
3.	Domestic multiple occupancy, places of assembly, medium to large factories and offices, main roads and railways.	3
4.	Essential services, valuable and/or costly property.	10
5.	Structures or services giving great danger if damaged.	30

standard statistical methods. This approach is most suited to flooding, where the likelihood of a flood of a particular magnitude is generally expressed by the **return period** or **recurrence interval**. Thus, the flood which is expected to be equalled or exceeded **on average** every 100 years, has a return period of 100 years. This event could occur any year, but the probability of its occurrence during 100 years is much greater than during a one year period.

The relationship between probability (expressed as a percentage), return period and the length of period under consideration is shown in Table 4.7. This indicates that a 1000 year event has a 6% chance of occurring during the lifetime of a building (taken here as 60 years).

Perhaps the single most important limitation on both deterministic and probabilistic approaches is the limited data sets of, for example, rainfall or wave climate records from which predictions have to be made. Benson (1960) demonstrated that to achieve 95% reliability on the estimate of discharge of a 50 year flood event required 110 years of records; such lengthy data sets are not common in Great Britain, highlighting the problems associated with defining the likelihood of extreme events. Detailed rainfall records are usually

available for 100–150 years, wave data are much scarcer (HR Wallingford, 1987) with extreme wave heights often extrapolated from relatively short data sets.

(vi) **calculating annual risk and reliability;** the specific risk associated with an event of particular magnitude can be calculated as follows:

$$\frac{\text{Likelihood of an Event}}{\text{Importance of Land Use} \quad \text{x} \quad \text{Vulnerability of Structure}}$$

The measure of chance that an engineering system or structure will not fail is known as the **reliability**. From an erosion and flooding perspective, it can be viewed as an indication of the ability of an area to support particular land uses or development without suffering major losses. Reliability is related to the probability of failure, i.e. risk, as follows:

Reliability = 1 − Risk

In many instances it is inappropriate to evaluate either risk or reliability in absolute terms because of the uncertainties in assigning values for the hazard and the assets at risk. In many cases it is more useful to assess the **relative risk** to each of the shortlisted sites from particular hazards, based on both factual data and subjective appraisal (see Chapter 5). The value of relative risk assessment is that it can quickly enable sites to be compared or allow decisions to be made about where limited resources and finances should be directed.

Table 4.6(a) Typical values of structural vulnerability to ground movement (after Cole et al, 1993).

TYPE NO.	TYPE OF STRUCTURE	VALUE
1	Structures designed to accommodate subsidence movements. Rigid frames. Ductile frames with an adequate number of possible plastic beam hinges.	0.7
2	Ductile coupled shear walls. Small building units with raft foundations.	0.7
3	Ductile frames with an inadequate number of possible plastic beam hinges. Small building units with strip foundations.	1.0
4	(a) Single ductile cantilever shear walls. (b) Garden walls.	1.4
5	Shear walls or slabs not designed for ductile flexural yielding but having the ability to sustain a significant amount of movement. Major electricity services within infrastructure. Roads.	1.4
6	Industrial processes which are sensitive to movement.	2.0
7	Buildings with diagonal bracing capable of plastic deformation in tension only: (a) Single storey. (b) Two or three stories. (c) More than three stories.	2.0 2.5 3.0
8	Single storey cantilevered structures: (a) Ductile columns of doubly reinforced load bearing walls providing restraint to the structure. (b) Boundary walls. (c) Single reinforced load bearing walls providing restraint to the structure.	2.0 2.0 2.5
9	(a) Shear walls or floor slabs other than as given above. (b) Chimneys, pipelines and tanks or reservoirs on the ground.	3.0

Table 4.6(b) Typical values of structural material vulnerability to ground movement (after Cole et al, 1993).

TYPE NO.	TYPE OF STRUCTURE	VALUE
1	Structural steel or flexible pavement construction, thermoplastics.	0.7
2	Structural timber: Shear–wall building Other buildings	0.7 1.0
3	Reinforced concrete.	1.0
4	Prestressed concrete (when used in elements which resist movements by flexural yielding).	1.4
5	Unreinforced masonry in panels exceeding 3m, cast iron, thermosetting plastics, glass.	2.0
6	Unreinforced masonry in panels exceeding 6m and buried service pipes with rigid joints and no relief bends.	3.0

(vii) **reviewing the likely attitudes to the levels of risk;** the degree of confidence of those likely to be exposed to hazard events can be as described subjectively from "totally unacceptable" to "of no concern", depending on the likelihood and the potential consequences of an event. Here, a distinction must be made between those risks that are tolerated voluntarily and involuntarily; it is widely recognised that

Table 4.7 Percentage probability of a particular magnitude event occurring in a particular period (after Ward, 1978).

Number of Years in Period	Average Return Period, in Years							
	5	10	20	50	100	200	500	1000
1	20	10	5	2	1	0.5	0.2	0.1
2	36	19	10	4	2	1	0.4	0.2
5	67	41	23	10	4	2	1	0.5
10	89	65	40	18	10	5	2	1
30	99	95	79	45	26	14	6	3
60	–	98	95	70	45	26	11	6
100	–	99.9	99.4	87	63	39	18	9
300	–	–	–	99.8	95	78	45	26
600	–	–	–	–	99.8	95	70	45
1000	–	–	–	–	–	99.3	87	68

Where no figure is inserted the percentage probability >99.9

individuals exposed to an involuntary risk may be considerably more wary of the consequences than if the risk was voluntarily accepted:

> "the individual exposed to an involuntary risk is fearful of the consequences, makes aversion his goal, and therefore demands a level for such involuntary risk exposure **as much as one thousand times less** than would be acceptable on a voluntary basis" (Starr et al, 1976).

Table 4.8 presents a matrix for relating attitudes to voluntary and involuntary risks of given likelihood events of different severity; this is based on the expressions presented in Cole (1993) and Cole et al. (1993). This table demonstrates how the attitudes to voluntary and involuntary risks differ more for greater severities of consequences. Cole (1993) provides a series of examples of the attitudes to reliability with respect to building over shallow mine workings. These are reproduced here, but modified to be relevant to coastal cliffs:

- **Total Loss;** in localities with a long history of ground movement, a loss of one house in a thousand (on average) each year ("moderate risk"; 99.9% reliability; Table 4.8 may be tolerated by the local authorities. The risk is voluntary but the authorities are **"circumspect"** ensuring that houses are on raft foundations and involve flexible construction. Usually families in affected areas are re-housed and houses demolished, without spreading alarm. Because of the continuing nature of the problem, management strategies are well developed. This scenario would be applicable in Ventnor, Isle of Wight.

In areas with more sporadic events the position is different. In Scarborough's South Bay, the occurrence of three major slides since the early 1700 suggests an annual likelihood of a major landslide similar to the 1993 Holbeck Hall failure of around 1 in 100. The slopes would have an annual reliability of 99% against complete failure and the local residents upon whom the risks are imposed involuntarily would be expected to find the level of risk as "not acceptable".

Table 4.8(a) Risk and reliability for total loss events (modified from Cole, 1993).

Level of Hazard	Annual Likelihood		Degree of Risk	Annual Reliability		Attitude to Degree of Risk	
	To Life	To Property		To Life	To Property	Voluntary	Involuntary
Very high	1 in 100	1 in 10	Very high	99%	90%	Very concerned	Totally unacceptable
High	1 in 1000	1 in 100	High	99.9%	99%	Concerned	Not acceptable
Moderate	1 in 10000	1 in 1000	Moderate	99.99%	99.9%	Circumspect	Very concerned
Slight	1 in 100000	1 in 10000	Slight	99.999%	99.99%	Of little concern	Concerned
Unlikely	1 in 1M	1 in 100000	Unlikely	100%	100%	Of no concern	Circumspect
Very unlikely	1 in 10M	1 in 1M	Very unlikely	100%	100%	Of no concern	Of little concern
Negligible	1 in 100M	1 in 10M	Negligible	100%	100%	Of no concern	Of no concern

Table 4.8(b) Risk and reliability for partial loss events (modified from Cole, 1993).

Level of Hazard	Annual Likelihood		Degree of Risk	Annual Reliability		Attitude to Degree of Risk	
	To Life	To Property		To Life	To Property	Voluntary	Involuntary
Very high	1 in 10	1 in 1	Very high	90%	50%	Very concerned	Not acceptable
High	1 in 100	1 in 10	High	99%	90%	Concerned	Very concerned
Moderate	1 in 1000	1 in 100	Moderate	99.9%	99%	Circumspect	Concerned
Slight	1 in 10000	1 in 1000	Slight	99.99%	99.9%	Of little concern	Circumspect
Unlikely	1 in 100000	1 in 10000	Unlikely	100%	100%	Of no concern	Of little concern
Very unlikely	1 in 1M	1 in 100000	Very unlikely	100%	100%	Of no concern	Of no concern
Negligible	1 in 10M	1 in 1M	Negligible	100%	100%	Of no concern	Of no concern

Table 4.8(c) Risk and reliability for minor loss events (modified from Cole, 1993).

Level of Hazard	Annual Likelihood		Degree of Risk	Annual Reliability		Attitude to Degree of Risk	
	To Life	To Property		To Life	To Property	Voluntary	Involuntary
Very high	1 in 1	10 in 1	Very high	50%	10%	Very concerned	Very concerned
High	1 in 10	1 in 1	High	90%	50%	Concerned	Concerned
Moderate	1 in 100	1 in 10	Moderate	99%	90%	Circumspect	Circumspect
Slight	1 in 1000	1 in 100	Slight	99.9%	99%	Of little concern	Of little concern
Unlikely	1 in 10000	1 in 1000	Unlikely	100%	100%	Of no concern	Of no concern
Very unlikely	1 in 1000000	1 in 10000	Very unlikely	100%	100%	Of no concern	Of no concern
Negligible	1 in 1M	1 in 1000000	Negligible	100%	100%	Of no concern	Of no concern

- **Partial Loss;** a highway authority may be responsible for a road crossing an unstable slope with an annual likelihood of damaging movement of 1 in 10 ("high risk", 90% reliability; Table 4.8). The authority, if aware of the circumstances, would be "concerned" about the risks, although they may, however, be aware that the road had no better reliability against damage caused by the effects of heavy traffic. In such circumstances the authority might install slope stabilisation measures, provide an early warning monitoring system or prepare contingency plans to divert traffic around potentially affected areas (eg. along the Isle of Wight Military Road; Barton and McInnes, 1988).

- **Minor Loss;** in some areas ground movement is very slow and intermittent and the associated property damage will be minor, but recurrent. If a number of damaging events can occur every year, the residents would most likely be "very concerned", with the slope reliability as low as 10%. If slope stabilisation measures are provided that reduce the likelihood of an event to 1 in 100, the attitude of the occupiers could change to "of no concern" as the reliability approaches 100%.

The availability of reliable records or basic earth science information can often present significant constraints to the degree of precision that a risk assessment can hope to achieve. At a broad scale (e.g. for a general assessment) it may be appropriate to restrict the assessment to a consideration of hazard potential. For less extensive areas relative risk assessment to make a comparison between sites is more appropriate. Detailed risk assessment, involving the precise quantification of risk and reliability will probably only be suitable for specific sites.

In England and Wales, the NRA is required to undertake surveys of flood prone areas (under the Water Resources Act 1991 S105(2)). The various NRA Regions are currently preparing **flood risk maps** which should identify the extent of land liable to flood, in relation to risk. There are no comparable formal arrangements for preparing maps of flood prone areas in Scotland. In these areas specific risk assessments may need to be undertaken by the local authority or developer.

Sediment Budget

The establishment of a sediment budget for a catchment or stretch of coastline will rarely be achievable without considerable investment in research from long term monitoring of sediment transfers to mathematical modelling of the sediment mobility. This should not, however, detract from the need for a clear statement of the significance of sediment supply and transport, especially where there is a need to consider the potential effects of mineral extraction. This is particularly so on the coast where sediment transport can be an important factor determining the sustainability of natural coastal defences, conservation sites and recreation areas. This less rigorous approach can frequently be undertaken through the interpretation of available evidence; the SCOPAC Sediment Transport Study (Bray et al, 1992) for the coast between Lyme Regis and Shoreham provides an excellent example for this type of approach. The key stages in such investigations are likely to include:

(i) **establishing a database** of consultants reports, unpublished theses, scientific literature, and records of sediments transport;

(ii) **establishing the distribution and character of the coastal erosion input** (Figure 4.4); in areas where the historical records of erosion are limited the following sources can provide an indication of the nature and significance of sediment supply:

- Historical maps to establish the cumulative land loss between different map editions;
- Aerial photographs to establish the nature of change between different survey dates;
- Geological maps and memoirs to identify the nature of the eroding materials.

(iii) **establishing the significance of offshore sediment sources;** the distribution of seabed sediments may be obtained from

Figure 4.4 The input of sediment into the coastal zone: a sediment transport model for West Dorset (after Bray, 1992).

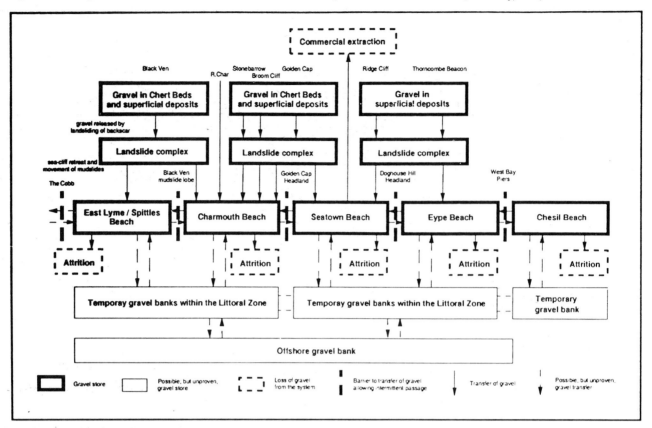

marine geological maps and memoirs or mineral assessment reports. Offshore bathymetry can be derived from Admiralty Charts. An indication of the mobility of seabed sediments may be available from existing reports and records. Alternatively it may be necessary to undertake a specific sediment mobility modelling exercise, involving:

- defining the wave and tide climate of the area of interest, using measured data and numerical model results;
- calculating the amount of time that seabed sediments would be mobile (Figures 4.5).

The SCOPAC and Crown Estates commissioned study of **South Coast Sea Bed Mobility** provides an excellent model for investigations of this nature (HR Wallingford, 1993b).

(iv) **identification of sediment transport pathways**; from available records or morphological evidence such as seabed gravel furrows or sand ridges, and the wave and tide climate, (Figure 4.6).

(v) **identification of sediment sinks**; marine geological maps or Admiralty Charts can be used to identify offshore sediment banks. Comparison between different chart editions may give an indication of the cumulative sediment accretion in these areas.

Permanent or temporary sediment sinks can be identified from a variety of sources, including geological maps, aerial photographs and available records of sediment transport. Comparison between different topographic map editions may give an indication of the cumulative sediment accreditation in these areas.

(vi) **identifying the nature and extent of significant sediment losses**; available records should be interpreted to establish losses from:

- marine aggregate extraction;
- dredging operations;
- offshore transport of sediment from features such as beaches, shingle ridges and sand dunes.

Figure 4.5 Seabed sediment mobility on the south coast (after HR Wallingford, 1993).

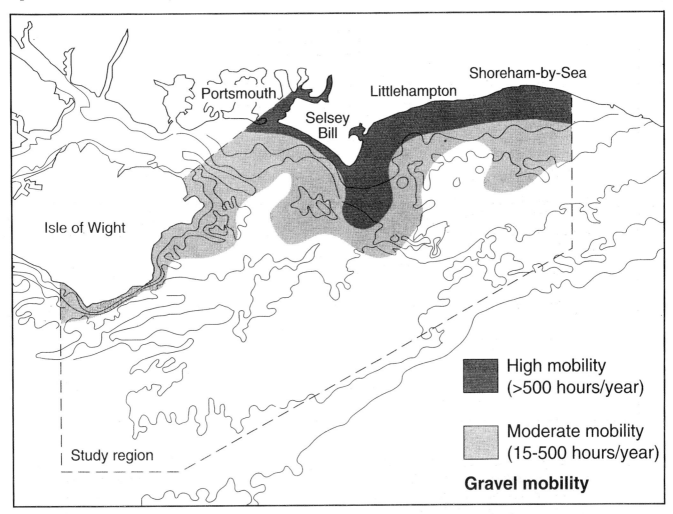

Figure 4.6 Principal sediment pathways along the south coast (after Bray et al, 1992).

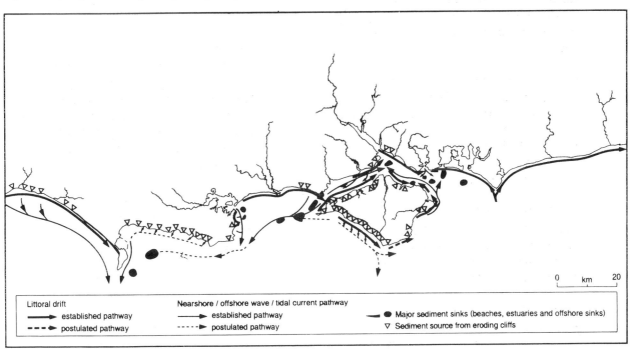

(vii) **establishing a sediment budget** by comparing the nature and volume of contemporary sediment **supply** with the sediment **demand** that is required to sustain coastal landforms.

Broad-based sediment transport studies of this nature are likely to form part of most shoreline management studies for littoral cells or sub-cells, and as such will be directly relevant to the **general assessments** needed by coastal planners at a strategic level. More detailed assessments involving the quantification of sediment transfers will follow similar principles, but will generally only be appropriate at a local level, when considering specific proposals for mineral workings or coastal defence works. In such circumstances the assessments would be the responsibility of the developer or operating authority.

Sensitivity Assessment

Precise assessments of the likely river channel or coastal changes that could accompany global warming or development can only be achieved through detailed research, involving long-term monitoring and sophisticated modelling. However, an assessment of the sensitivity of the various landforms in an area can provide an indication of where potential changes could occur and the significance of these changes to planners and managers. Such assessments are likely to involve:

(i) **establishing the potential for change** to the various landforms in a catchment or along a coast. This will involve an appreciation of:

- the origin of the various landforms (some may be relic features, created under former environmental conditions e.g. some shingle ridges);
- the nature and pattern of historical change;
- the factors influencing the pattern of change;
- the causes and mechanisms of change;
- the dependency of the landforms on continued erosion or supply of sediment.

(ii) **assessing the nature of the potential change**; a number of approaches may provide useful information:

- analysis of historical records;
- assessment of similar landforms currently experiencing different erosional or depositional conditions. For example, the degradation of abandoned sea cliffs can be used as an analogy for the effects of coast protection works on cliff conditions. Alternatively, similar landforms on more exposed coastlines can be studies to assess the possible consequences of increasing the rate of erosion;
- prediction of conditions at unmonitored locations from known conditions at monitored sites.

(iii) **assessing the potential consequences of change** which can be considered as part of a 4-fold classification:

- **dramatic change**; major, permanent change, including shifting of river channels or harbour mouths, etc;
- **significant change**; marked increases or decreases in, for example, the rate of erosion, landslide activity or frequency of flooding;
- **minor change**; examples include localised expansion or contraction of the areas at risk from erosion, deposition and flooding processes;
- **no significant change**.

(iv) **assessing the relative sensitivity of landforms**; individual landforms can be ranked according to the potential for change and the significance of these changes. In preparing a map of sensitivity it may be useful to adopt a qualitative scale (e.g. very low, low, moderate, high and very high).

Many shoreline management plans will have evaluated future trends of coastal changes within littoral cells or sub-cells; these studies will be of direct relevance to the preparation of general assessments and strategic planning. Further detailed information would generally be necessary when considering specific proposals on sensitive coastlines; this would need to be carried out by developers or operating authorities.

Resources and Constraints Maps for Planners

There can be little doubt that the final output of any earth science data collection and interpretation exercise should be structured in terms of planning and management issues as this will be the starting point taken by planners when considering any area. In most cases it will be appropriate to compile summary maps and descriptions of resources and constraints:

(i)   **resources**; these may include:

- landscape designations;
- nature and geological conservation designations;
- high grade agricultural land;
- mineral resources;
- water resources;
- ancient monuments.

In many instances it may be appropriate to identify the extent to which the management and utilisation of these resources is influenced by the continued operation of natural processes. For example, many geological SSSIs are dependent on fluvial or marine erosion for maintaining their scientific value.

(ii)   **constraints**; these may include:

- problem ground conditions;
- flood and erosion risk areas;
- natural coastal defences dependent on continued sediment supply.

These summary maps should generally be accompanied by a concise, jargon-free, report which highlights the significance of the various resources and constraints for planning and development. This report can also provide a clear explanation of the relevant key considerations.

Scope of Investigations

Although the general procedures of inventory, interpretation and summary are applicable at each of the various levels of investigation (general assessment, site reconnaissance and detailed assessment), there will be considerable differences

in the required precision and reliability of the results and, hence, the quantities of data that will need to be collected and the scale of mapping. This will, of course, have significant implications for the nature and scope of such investigations, as will be outlined in the following Chapters 5–7.

It is important to stress that the land use character of a catchment or coastline and the extent to which there is significant development or redevelopment pressure will have a significant influence on the nature and amount of earth science information needed to support decision–making. Detailed assessment and site reconnaissance will be required most frequently in considering allocations of land for development and in **developed areas,** although such studies will have to take account of the relationship between the site, the proposed development and the surrounding countryside or coastline. General assessments will need to be undertaken for the **whole of an administrative area** as such studies provide the background information on the principal physical characteristics of an area for both strategic and local planning, i.e. there is a need for a comprehensive planning base for the whole area so that the full range of options for development and conservation can be considered.

It should be recognised that the effort that may need to be expended will be dictated by the nature of the land use. The enormous variety of settings makes it wrong to be dogmatic, but the following points may provide a useful guide:

- **risk assessment**; in developed areas it will be necessary to take account of the problems that may occur, how frequently events may take place and whether they can be satisfactorily overcome by proposed developments or uses. In undeveloped areas it will generally be sufficient to identify the distribution of vulnerable areas;

- **sediment budgets**; on developed coasts, especially those with potential mineral resources, it will be necessary to address the significance of particular sediment sources and sinks and establish the nature of the principal sediment pathways. On the undeveloped coast and most river catchments, awareness of the general pattern of sediment supply and transport will probably be adequate;

- **sensitivity**; on the developed coast it will be necessary to have a clear understanding of the likely consequences of, for example, sea level rise, whereas on the undeveloped coast and most river catchments a broad appreciation of the possible changes to existing trends may be appropriate.

The information that is needed to address planning and management issues at different scales will generally be obtained in a variety of ways:

- information specially compiled by various organisations for other aspects of management, including monitoring programmes, shoreline management studies, flood risk maps, estuary management plans, SSSI management plans, etc;

- information previously collected as part of national mapping programmes or data collation exercises that cover both inland and coastal areas. Examples include British Geological Survey Geological Maps and the National Landslide Databank;

- information that can be **derived** from readily available sources. Examples include the measurement of land loss through erosion from historical sequences of topographic maps or delineation of flood prone areas from detailed Ordnance Survey Maps;

- information that is not currently available but will need to be collected as part of the investigation of specific sites or areas.

The information sources relevant at different levels of investigation will be discussed in Chapters 5 – 8. However, it is important to stress that both planners and developers should try to make maximum use of the information held by the various bodies with an interest or responsibility in environmental management. Indeed, planners should endeavour to make full use of the experience available within different departments of their authority, or within bodies such as the NRA and conservation agencies before undertaking new surveys or investigations. Where new surveys are required, local planning authorities should seek advice from these departments and bodies who may be able to provide appropriate expertise to assist in the design of the investigation and interpretation of the results. In some circumstances it may be appropriate to use commercial consultants to advise on particular aspects of catchment or coastal processes.

It is important to ensure that the effort spent is appropriate to the level of investigation. In many instances it is not feasible to collect and collate all existing data relevant to any study, partly because some holders will not provide it and partly because considerable time may be needed to access widely dispersed data sets. Often planners and developers will need prompt access to general guidance rather than undertaking investigations which may involve many years work before detailed results become available. There comes a points beyond which investigation is no longer cost–effective in terms of the amount of additional material they retrieve and the overall value of the information to the final product. Limitations to the time and resources available will generally dictate that comprehensive geographical or thematic coverage is unlikely to be a realistic option. Investigations will, therefore, be a practical compromise addressing **key issues** that relate to specific policy needs, with effort directed towards areas where there is significant pressure for development.

Chapter 4: References

Barton M E and McInnes R G 1988. Experience with a tiltmeter–based early warning system on the Isle of Wight. In C Bonnard (ed) Landslides, 379–382. Balkema.

Benson M A 1960. Characteristics of frequency curves based on a theoretical 1000 year record. USGS Water Supply Paper 1543–A.

Bray M J 1992. Coastal sediment supply and transport model for Dorset. In R J Allison (ed) The Coastal Landforms of the Dorset Coast. Geologists' Association Guide No. 47, 94–105.

Bray M J, Carter D J and Hooke J M 1992. Coastal sediment transport study. Report to SCOPAC. Dept. of Geography. University of Portsmouth.

British Geological Survey 1994. Catalogue of Maps. BGS, Keyworth.

Bromhead E N 1986. The Stability of Slopes. Surrey University Press.

Cole K W 1993. Building over shallow mines. Paper 1: Considerations of Risk and Reliability. Ground Engineering, Jan/Feb. 1993.

Cole K W, Jarvis S T and Turner A J 1993. To treat or not to treat abandoned mine workings: towards achieving a dialogue over risk and reliability. In B O Skipp (ed) Risk and Reliability in Ground Engineering, 1–28, Thomas Telford.

HR Wallingford various dates. Macro Review of the Coastline of England and Wales. 10 Volumes.

HR Wallingford 1987. Wave data around the coast of England and Wales. A review of instrumentally recorded information. Report SR113.

HR Wallingford 1993. South Coast Seabed Mobility Study. Report EX2795.

MAFF 1993. Strategy for flood and coastal defence in England and Wales. MAFF Publications.

MAFF 1994. Shoreline management plans. A guide for operating authorities. MAFF Publications.

NERC 1975. Flood Studies Report. Institute of Hydrology, Wallingford.

Rendel Geotechnics 1993. Coastal planning and management: a review. HMSO.

Rendel Geotechnics 1995. The occurrence and significance of erosion, flooding and deposition. Report to DoE. Erosion, planning and management: a review. HMSO.

Ritchie W and Mather A S 1985. The Beaches of Scotland. Countryside Commission for Scotland.

Varnes D J 1984. Landslide hazard zonation: a review of principles and practice. UNESCO.

Ward R C 1978. Floods: a geographical perspective. MacMillan Press.

50

5 General Assessments

Introduction

General background guidance on the nature of physical conditions in an area is often needed to ensure that the issues associated with erosion, deposition and flooding are fully appreciated by decision makers. Failure to recognise potential problems at an early stage of the decision making process can lead to increased development costs, abandoned development, calls for publicly funded defence works, damage to adjacent land and property, an increase in the risks to other land users or a degradation of environmental resources or a degradation of environmental resources.

For many purposes, such as planning policy formulation and preparing a regional context for catchment or shoreline management, a broad brush approach, which highlights the physical conditions and related issues which need to be borne in mind, can contribute significantly to the safe, cost–effective development and use of land. **General assessment** of an area can provide a relatively quick appraisal of conditions over large areas by collecting and interpreting readily available data sources. Such information can be used to:

- update and improve national policy or strategic advice;

- prepare strategic policies for addressing issues associated with erosion, deposition and flooding e.g. management of hazards, safeguarding of natural resources etc.;

- prepare detailed policies for addressing erosion, deposition and flooding issues in areas where the risks to existing or future development are readily managed;

- provide general guidance on whether development in particular areas could have adverse effects on the level of risk elsewhere or on the physical environment (e.g. conservation interests, mineral resources).

Basic information can be compiled into an **inventory** map of catchment or coastal geomorphology, highlighting some or all of the following features: geology, topography, man-made and natural flood and coastal defences, sediment supply and transport pathways, the nature and trend of change. The result of this compilation should form the basis of the subsequent assessments of **risks**, **sediment budget** and **sensitivity**. Deficiencies in the existing data sources for a particular area will highlight the need for specifically commissioned data collection exercises. For most coastal studies, shoreline management plans will provide much of the information necessary for this level of investigation.

Sources of Information

In areas where erosion, deposition and flooding are likely to impose **constraints** to development and land use, decision makers will need to consider identifying those areas where particular consideration should be given to these issues. They may also need to be aware of the type of problems that may occur, how frequently that damaging events may take place and whether they can be satisfactorily overcome by proposed developments or uses. This requires access to information on:

- the distribution of vulnerable areas;
- records of past events;
- the likelihood of damaging events;
- the possible effects of development.

Decisions makers will also need to be aware as to whether conservation features or recreational, resources such as beaches and sand dunes, are dependant on the continued operation of natural processes. This requires information on:

- the distribution of potential resources;
- the sensitivity of these resources to the effects of development.

Lack of awareness of the potential for damaging events in an area is often a significant obstacle to be overcome, especially where the threat is associated with extremely rare events (e.g. flash flooding in dry valleys and tributary streams) of failure of man-made structures (e.g. dams and embankments). It is important, therefore, that the first step in any general assessment should be to recognise that problems can be expected to occur in most administrative areas (Figure 4.3). For most authorities the key problems will relate to flooding and, on the coast, cliff erosion and landsliding; the principal sources of general information for addressing these problems will, therefore, be outlined in the following sections.

Surveys of Flood Area

The **Water Act 1973** S.24(5) required the then Water Authorities in England and Wales to carry out surveys of their areas in relation to their land drainage functions. These surveys, known as Section 24(5) surveys, were designed to identify, map and tabulate data on areas that might benefit from flood protection or land drainage. Following the privatisation of the water industry in 1989, the NRA were made responsible for undertaking surveys of flood prone areas (the **Water Resources Act 1991 S105(2)**). The various NRA regions are currently in the process of updating and revising the flood risk maps inherited from the Water Authorities. This process is being influenced by the requirements imposed by recent Government advice on flood risk (Development and Flood Risk; DoE Circular 30/92 – MAFF Circular FD 1/92 – Welsh Office Circular 68/92) which states that the surveys should:

- indicate the areas where flood defence problems are likely;

- identify the extent of floodplains, washlands and other land liable to flood, in relation to risk.

It would appear likely, therefore, that the revised NRA flood maps (known as S.105 maps) will show the extent of flood prone land at a scale suitable

for use by local planning authorities and the risk of flooding in relation to the **standard of protection** which exists or is to be provided (see Chapter 11). The NRA has recently published a "Memorandum of Agreement" with local planning authorities on the subject of S.105 maps. The aim of this Memorandum is to set a framework for the NRA to discuss with planning authorities what flood risk information is required as an input to development plans and where development pressures are such that the NRA can target its resources. Indeed, it is recognised that a complete set of S.105 maps is not immediately achievable and the Memorandum seeks to ensure that key areas are covered first, within the resource constraints of the NRA, (Pickles and Woolhouse, 1994).

In Scotland, there are no comparable formal arrangements for preparing maps of flood prone areas. Information, albeit to varying levels of reliability and availability, will be held by various departments within the Island and Regional Councils and the River Purification Boards (Table 5.1). In general, basic information is generally only available where particular departments have a specific requirement to maintain records and at a level of detail which reflects the perceived significance of flood problems within an authority area.

Recession of Soft Cliffs

Although sub aerial processes can be locally important, most erosion of cliffed coastlines is achieved by cliff falls and landsliding and is usually stimulated by wave attack. Indeed, all cliffed coastlines are testimony to the cumulative efficacy of landslides (the term is used here to describe the full range of mass movement forms, from falls to slides and flows, and including seepage erosion), for they are, in reality, the coalescent scars of innumerable individual failures. However, not all clifflines are undergoing rapid rates of retreat. Those formed of harder rocks underlying some areas of Britain are retreating very slowly so that landsliding is of limited frequency, and often relatively small scale. By contrast, where failures are more frequent and sometime of a larger scale, the rates of cliff recession are higher; many of the most dramatic rates occur on cliffs developed in soft sedimentary rocks and of weak glacial deposits that form the margins of lowland Britain.

Table 5.1 Sources of information on flood risk in Scotland.

• Regional Drainage Dept;	Information confined to urban areas and associated with the implementation of the Flood Prevention (Scotland) Act 1961.
• River Purification Board;	Hydrometric data covering rural areas, but little detail of flood levels or the extent of flooding is documented.
• Regional Roads Dept;	Only flooding associated with roads or other public property.

The coastline of Great Britain is very long and well known for its varied character, rapidly changing rock type and local intensity of marine erosion, so it should come as no surprise that the major areas of coastal slope failure are correspondingly diverse. Four broad types of coastal recession can be recognised on the basis of the nature of ground-forming materials and style of landslide activity (Hutchinson, 1984; Jones and Lee, 1994).

i. **Major coastal landslides in weak superficial deposits**; The east coast of England from Flamborough Head to Essex and parts of North Yorkshire is largely developed in thick sequences of glacial till interbedded with sands and gravels. These deposits can be rapidly eroded by the sea; for example, the entire 60km length of the undefended Holderness coastline (Humberside) has retreated at rates of 1–6m pa since 1852.

ii. **Major coastal landslides developed in stiff clay**; Stiff clays are particularly prone to landsliding, with class examples occurring along the southern shore of the Thames estuary in Kent, where cliffs up to 40m high developed in London Clay have repeatedly failed in response to marine erosion, which results in average retreat rates of up to 2m per year.

Although minor failures occur elsewhere along the coast where clay strata outcrop, the largest failures are associated with interbedded clays and sands. There are conspicuous failures of this type on the north coast of the Isle of Wight, especially at Bouldner. At Barton-on-Sea in Christchurch Bay landslides extend for 5km on cliffs up to 30m high, developed in Barton Clay and Barton Sand overlain by Plateau Gravel.

iii. **Landslides developed in stiff clay with a hard cap-rock**; The largest coastal landslides occur in situations where a thick clay stratum is overlain by a rigid cap-rock of sandstone or limestone, or sandwiched between two such layers. Amongst the most dramatic examples is Folkestone Warren, Kent, where the high Chalk Cliffs have failed on the underlying Gault Clay, the Isle of Wight Undercliff and on the West Dorset coast at Black Ven.

iv. **Landslides developed in hard rock**; Coastal cliffs developed in rocks are continually suffering minor collapse due to basal undermining by the sea. These events are most frequent in the soft rock clifflines of south-eastern England, such as the famous Seven Sisters and Beachy Head chalk cliffs of East Sussex, which are currently retreating at an average rate of 0.97m a year. Large falls also occur on a number of coasts including the Triassic sandstone cliffs of Sidmouth, Devon and the Liassic limestone cliffs of South Glamorgan.

MAFF have recently undertaken a survey of coast protection defences in England to examine their extent, adequacy and state of repair (MAFF, 1994). The survey also provides information relating to unprotected and eroding cliffs, identifying 860km of significantly eroding coastline. In this context, the Coast Protection Survey can provide valuable strategic data to ensure that coastal defence policies are developed that are compatible with land use and conservation interests.

The most readily accessible sources of information about coastal cliff recession are Ordnance Survey maps and historical maps (see Table 4.2). Charting cliff retreat between maps of different dates might appear straightforward, but in fact such sources can frequently be misleading. Early maps are often unreliable in the depiction of cliffs in undeveloped areas; the accurate representation of relief and landshape in areas of rapid erosion or active landslide complexes would not have been a high priority for the early Ordnance Survey surveyors and their predecessors. Indeed, the depiction of steep slopes by means of hachuring often owes more to artistic licence than to an accurate representation of topographic form. In some instances, there may also be significant differences in **map projection** between sets of historical maps.

Analysis of historical maps can give an indication of the average annual rate of erosion along a particular stretch of coast. However, cliffs do not retreat in a uniform manner; long periods of relative inactivity are separated by infrequent large events. For example, during the 1953 east coast storm surge, glacial sand cliffs at Covehithe retreated over 13m overnight, in comparison with long–term average rates of around 5m per year.

Vertical aerial photographs provide a more reliable source of information about cliff erosion rates, although they are generally only available from around 1940 onwards. In contrast to many maps, photographs allow clear identification of the evolution of coastal landslide systems as well as the accurate determination of the pattern of cliff retreat. They do, however, suffer from the same limitations as maps in that they are incidental observations made at a fixed time and, hence, tend to only give limited indications of the nature and rate of contemporary processes.

Measurement of recent and current erosion rates can be used to define an area which is likely to be affected by instability or erosion in future decades. These estimates can be improved through an appreciation of the nature and scale of individual landslide events that could be expected on different stretches of cliff. In this context, it is worth noting that MAFF have recently established a programme of research to develop methods of predicting soft cliff recession rates suitable for both detailed and strategic planning. The initial findings from this study indicate that it is likely that the most effective approach to prediction will involve an hierarchy of complementary techniques:

i.   **historical assessment**; suitable for measuring past recession rates over large areas where only a general indication of the future trend is required. This should utilise historical map sources, sequential aerial photographs and incidental records, although each source has particular limitations and a number of factors may

restrict the validity of projecting historical rates into the future (Table 5.2);

ii.   **geomorphological assessment**; suitable for assessing future recession rates over relatively large lengths of coast, particularly for use in defining set–back lines or determining general sediment budgets in littoral cells. A key principle of this approach is that cliffs of similar materials and structure will tend to retreat through similar mechanisms. Here, the **potential** for cliff recession can be expressed as:

- the range of cliff failure mechanisms that could occur in a particular cliff environment;

- the scale of individual events of different mechanisms.

Mapping the landslide features along a particular geological outcrop can provide important indications of recession **potential**, as there are frequently examples of a wide range of scales of landslide forms. In this context, it is significant that the rarest and largest landslide features tend to be the most **persistent**, i.e. they can remain visible in the landscape for long time periods. The geomorphological mapping needs to be supported by **historical assessment** to give an indication of the frequency of events (from incidental records) and the cumulative loss of land (from maps and photos). A model can then be developed which:

- indicates how many events could have occurred over a particular period, i.e. how many cliff recession **episodes**. This can be related to the frequency of causal factors such as rainfall or wave climate to give an indication of the potential frequency of **formative events**;

- highlight a range of possible scenarios for future behaviour, based on the potential for events of various sizes and the frequency of formative events;

Table 5.2 A summary of the possible limitations of using historical rates to predict cliff retreat.

- Historic recession rates may not show a consistent trend, due to insufficient data or due to highly irregular recession rates. Statistical tests will be needed to identify whether a significant trend can be attributed to the data.

- If a consistent trend can be identified, then this can provide a basis for predicting future recession rates. However, there is likely to be scatter in recorded recession rates, and methods will be needed to extrapolate from the data to predict future rates.

- Conditions may not be stationary, due to changes in environmental loading (eg sea level rise, increased storminess). These effects will not be identifiable in historic data on recession rates. If these changes are significant, then confidence in prediction will be reduced, and predictions may need to be modified to account for future changes in loading.

- There may be great variability in the geology which can have a significant effect on future recession rates, and invalidate predictions from historic data.

- indicate the likelihood of events of specific magnitude and the relative risk of different cliff–top losses in a particular event.

iii. **empirical models**; suitable for assessing recession rates over short distances of eroding cliffs in particularly vulnerable locations to identify relative risk and prioritising the allocation of resources for coast protection. Such models need to be based on relationships established between key variables on cliffs (e.g. height, material, structure), the shoreline (e.g. shore platform angle, beach volume etc.) and the marine environment (e.g. exposure, wave climate etc.).

iv. **probabilistic process models**; suitable for predicting recession rates at specific sites, especially in high risk areas, where recession models are required to evaluate the effectiveness of a range of potential erosion control techniques.

Coastal Landsliding

In 1984 the Department of the Environment commissioned a national review of the extent of landsliding throughout Great Britain (inland as well as on the coast), which formed part of a strategy for addressing the need for greater consideration of the potential problems by the planning system (Geomorphological Services Limited, 1986/87). The study was a census of reported landslides which revealed a grand total of 8835 as shown on the distribution map (Figure 5.1); with 6120 were recorded in England of which 1013 are coastal,

1200 in Scotland (175 on the coast) and 1515 in Wales (114 on the coast).

As a result of the review, distribution maps of landslides recorded in the **published literature** have been produced for all counties in England and Wales and for Scotland. The maps are available at a scale of 1:250,000 and show the distribution of landslides recorded in the original census which was completed in 1985. Each recorded landslide is portrayed with a unique reference number and is described in terms of landslide type and age. In addition, areas of suspected but unrecorded instability are shown (Figure 5.2).

To encourage the wide use of the survey results the information has been compiled in a computerised archive – the **National Landslide Databank**. This database includes the results of the original census and also landslides identified in an updating programme that was completed in 1991. Landslide information for an individual county or Scottish region can be purchased on computer discs for direct access using an IBM PC or fully compatible desk top computer. Alternatively, a telephone enquiry service is available for identifying recorded landslides within a specified area. The charge covers the cost of providing:

- location data;
- database listing for each landslide record;
- relevant bibliographic references.

The National Landslide Databank and associated maps can provide a general indication of where coastal cliff erosion problems are likely to occur. Further information and the nature and cause of potential problems can be obtained by accessing

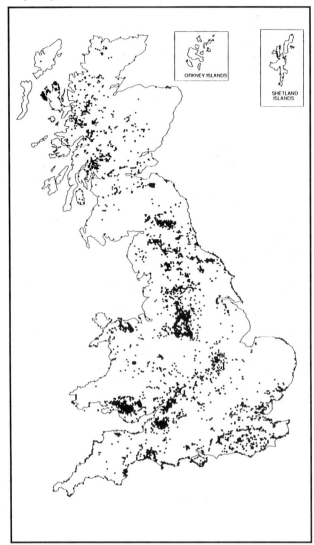

the original source material highlighted by the bibliography attached to the database.

Although the Databank represents the most complete data set on the nature, occurrence and extent of landsliding in Great Britain, it is wholly composed of reported occurrences and is, therefore, an artifact of knowledge rather than a true representation of the actual distribution of coastal cliff erosion problems. It must be anticipated that in many areas the information that can be obtained from the Databank will need to be supplemented by other sources to gain a realistic appreciation of the actual pattern of landsliding. Probably the most cost−effective information source are vertical aerial photographs available from the Ordnance Survey or commercial air survey firms. Coastal landslides can be identified from aerial photographs because of their distinctive surface form and tonal patterns (Table 5.3). The key to successful interpretation is

a systematic search of the coastal cliffs where instability may be anticipated. The most appropriate photograph scale is probably around 1:25,000, although larger scales provide more detail they will require considerably more effort to inspect large areas.

Records of Past Events

It is particularly important for decision makers to be aware of the historical incidents of damaging events in an area as this will provide a general indication of the nature and scale of potential future problems and, by inference, the level of risk than could be expected. Table 5.4 highlights the range of sources that could provide valuable background information. Some are more accessible than others, some may only need to be searched where there is a pressing need to assess potential problems at a particular site (see Chapters 6 and 7). In many areas the collation of historical records will have been carried out by the NRA or local authority.

The records held by the various operating authorities (e.g. the NRA, coast protection authorities etc.) **will, in many instances, be the most valuable sources of information;** it is important, therefore, that these bodies are consulted at an early stage of an investigation. It is important to bear in mind, however, that documents such as journals and diaries can include valuable descriptions of floods and coastal landslide events. For example, in 1811 Thomas Webster wrote to Sir Henry Englefield about a major landslide at the eastern end of the Isle of Wight Undercliff:

".... I now saw this section extending for many miles, appearing like a huge wall sheltering the land from the north. On the top of it were several low hills of marl; whilst, below, the whole of the country between it and the sea, had, apparently, at some remote period, been one immense mass of ruin, though now covered with woods, cornfields, and villas. It appeared that the foundation upon which this part of the island rests had given way towards the sea, parting along the line which now forms the face of the cliff. Prodigious masses of the strata had fallen, or rather slid, into the ocean, where they now remain in the inclined position into which they originally fell, chiefly dipping towards the land; whilst smaller portions stop short,

Figure 5.2 A sample portion of a 1:250,000 scale National Landslide Distribution Map.

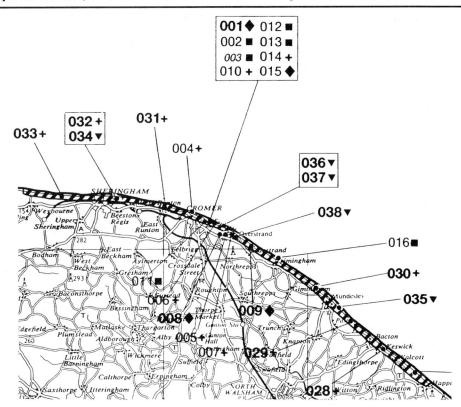

Distribution of Landslides

- • Location of landslide (<0.25 km^2)

- 🖤 Extent of landslide (>0.25 km^2)

- ▨ Landsliding of uncertain extent and character

- ▦ Area of numerous landslides, too concentrated to map individually

- ▨ Unstable screes

- ▥ Cambered/ Foundered strata (◇ <0.25km^2)

- ▤ Possible areas with extensive ancient landsliding

Type of Landslide (Divisions by material type not shown)

+ Unclassified	**Slides (Rotational)**	
◆ Complex	■ Single	
Falls	□ Multiple	
∗ Fall	▣ Successive	
○ Topple	**Slides (Translational)**	
⊙ Spreading or sagging	▲ Non-rotational (Block glide: Slab slide)	
	△ Planar (Rock slide: Debris slide: Mudslide)	

Flows

- ▼ Rock avalanche: Debris flow: Mudflow: Bogburst

Landslide Reference Numbering System

(27) County number

017 Slide number

27017 Landslide reference number

Landslide Age (represented by size and style of number)

017 Active (currently unstable or cyclically active with period up to 5 years)

017 Recent (movement within last 100 years)

Table 5.3 Typical indicators of slope instability from photographs.

PROCESS	KEYS TO RECOGNITION ON AERIAL PHOTOGRAPHS
Rockfall	● light striations of freshly exposed rock
	● stripped vegetation
	● cones of light coloured fresh scree
Rotational failure	● arcuate backscar above concave slump feature
	● intact strata exposed in failure scar
	● depression behind back-tilted strata, possibly occupied by a pond or lake
	● oval or lens shaped areas of cultivation
	● walls destroyed, trees curved at base or leaning back
	● rivers displaced abruptly
	● cliff forms change suddenly
Translational or planar side	● geological outcrops dipping in same direction as ground
	● ruckled or uneven ground in front of step feature
	● angular backscars or crevices following joints
Debris slide	● light coloured areas with flow patterns
	● road re-alignments
Rotational slump	● arcuate backscars above concave slump feature
	● dark tones over damp or waterlogged ground
	● lobate toe at advancing edge of slump
Debris and mud flow	● hummocky, stepped or fissured ground avoided by cultivation
	● displaced boundaries, disrupted vegetation patterns
	● patterns of light and dark tones
	● isolated boulders remote from outcrop
	● debris outwash fans on valley side
	● cones of loose, unvegetated scree

and lay dispersed in all directions, the intervals being filled up with chalk, marl, and other substances from above ..... The enormous masses of rock, forming by their pressure new and firmer foundations, seem ever since to have remained unmoved." (Webster, in Englefield, 1816).

This documented example, and others, were used to establish the contemporary ground behaviour of the coastal landslide complex at Ventnor, Isle of Wight (Lee and Moore, 1991). Over 200 individual incidents of ground movement and coastal erosion were identified, allowing a detailed model of landslip potential to be developed that formed the basis of an integrated approach to planning and management within the area (Lee et al, 1991).

Local newspapers are a very important source of information about significant flood and coastal erosion events over the last hundred years or so. For example, the major floods of the 19th century, such as those of October 1875 were described in considerable detail, sometimes occupying almost entire newspapers. Often the accounts may contain references to previous similar events. As an example the following summarises an article appearing in the 'Nottingham Daily Guardian' for January 2, 1901:

"A heavy and prolonged downpour on Sunday, 30 December 1900. It was recorded locally as 1.684 inches, and compared with the year's previous heavy fall of 1.286 inches on June 11 1900. The water–level rose rapidly during the night of Monday 1 Jan/Tuesday 2 Jan 1901, and by the morning of the 2nd flooding was taking place over both banks of the river. By Tuesday night the flood level had passed the 1869 flood mark and had almost reached the 1852 flood mark."

The process of determining relative flood height can be carried back through the 19th century and well into the 18th, depending on the numbers and longevity of newspapers themselves. Their value is enhanced by the fact that they enable a systematic review of erosion, deposition and flood events over long periods. It is important to bear in mind, however, that historical reports, like current ones, tend to concentrate on the events that affect people or property and, hence, most records relate to built up areas. When researching particular events that have been recorded in local newspapers or documents, it is necessary to make a judgement on the reliability of the data source. Potter (1978) suggests that three questions need to be borne in mind:

- what is the nature of the event being recorded, and with what detail, and is it pertinent to the stated objectives?

- who is making the report, in particular what are his qualifications to know of the event, i.e. is it a personal observation based on his own experience; an editing of reports from other people, who themselves may have edited the information; a plausible rumour; or a complete invention; or falsification?

- in the light of knowledge of this type of event, is the report credible, in whole, in part, or not at all?

Events discovered by historic search generally tend to be more subjective, and usually contain much less detail than recent reports. However, even a present day flood or landslide report contains edited information. It should be remembered that most were originally recorded for purposes other than recording the erosion, deposition and flooding aspects of the events and its circumstances.

The Likelihood of Damaging Events

In areas which have been identified as being vulnerable to erosion, deposition and flooding events, decision makers will need to make judgements as the level of risk that may be acceptable. At a general level this involves establishing how frequently an event of a given size could be expected. This may involve determining, for example:

- the water level associated with a flood of a specified return period;

- the rate of coastal erosion that may occur over a specified time period.

In England and Wales, the estimated return periods of major flood events for a particular catchment or low lying coastal area can be sought from the NRA. As already described, the flood survey maps produced by the Water Authorities may provide an indication of the potential scale of problems associated with a specific return period event. The updated maps, to be produced by the NRA, are required to show the level of risk in flood areas, probably related to the standard of protection which exists or is to be provided. This information could be used to indicate the areas that could be affected by a particular event, or the depth of flooding than could be expected at a particular site during such an event.

In areas of significant coastal erosion, there may be a need for decision makers to have background information about the recent and current rates of cliff recession. This, for example, could help ensure that development does not take place in areas where instability is likely to occur during the lifetime of the building. In many instances this may be available from the Engineering or Technical departments of maritime district councils.

The Distribution of Potential Resources

Maps showing the location of various statutory and non–statutory conservation designations can be obtained from the relevant country conservation agency (English Nature, CCW, SNH), who can also supply site descriptions detailing the conservation value of particular features of interest.

Table 5.4 Historic sources of flood and coastal erosion event information from the 16th century to present day (from Potter, 1978).

PERIOD	SOURCE	COMMENT
Present day and recent past	NRA Records Coast Protection Authority Records	The NRA, the former Water Authorities and their predecessors would usually produce a report on major flood events. Information may be contained in these reports on: – flood cause and mechanism – estimated discharge – impact If no reports have been prepared, relevant flow and level records may be available. Coast protection authorities are likely to hold records of damaging events, especially in urban and defended areas. In Scotland, RPAs and Regional Councils may have prepared reports on major events.
	Newspapers	National and local newspapers will probably have accounts of significant events, including photographs of damage, etc.
	Rainfall statistics	Information on the climatic events which have caused flooding or coastal erosion can be found in a variety of journals, including: – British Rainfall (1860–present) – Meteorological Magazine (1866–present)
	Surface Water Year Book	Provides a monthly summary of discharges of British rivers.
18th and 19th Centuries	Newspapers	The British Museum Newspaper Collection held at Colindale, London contains all provincial newspapers and London newspapers since 1801. Articles may contain information on: – flood height – extent of inundation – type of cliff failure – damage "Gentleman's Magazine" dates from 1731 and includes information on flood events and notable landslides.
	Local Histories	Local histories are available in public reference libraries or other collections; they may contain information on damaging events.
	Directories	Many counties are the subject of Directories with special sections on the larger settlements. Flood and weather phenomena appear in some; for example, a Lincolnshire Directory (White, 1881) contained a list of storm surge heights recorded at Boston between 1791 and 1877. Norton (1950) provides an index of the availability of Directories for particular areas.

Table 5.4 (cont ...)

PERIOD	SOURCE	COMMENT
18th and 19th Centuries (continued)	Specialist Sources	Examples include: – County Archivist's Offices – Borough records – Town Council records The **Royal Commission on Coast Erosion**, 1906 and the **Royal Commission on Land Drainage**, 1927 contain detailed replies from many local authorities on the problems encountered in their areas.
16th and 17th Centuries	Public Records Office	The vast national collections began to be catalogued in the 19th century. The **Calender of State Domestic** cover the period 1640-1704, and may include details of floods and weather-related events. The **Acts of the Privy Council** provide similar coverage for the period 1542-1631. The nature and extent of the Public Records Office collection is contained in the "Guide to the Contents of the Public Record Office", (1963-1968).
	Diaries and Chronicles	These are a well recognised source and many have been reprinted (see Matthews, 1950 for a coverage of British Diaries 1442 -1942). Famous example include Samuel Pepys (1660-1669) and John Evelyn (1620-1706) for London; Anthony à Wood who provides an almost complete calendar of weather and floods for the 17th century.
	Specialist Sources	Examples include: – Sewer Commissioners and Courts of Sewers records – Ecclesiastical records

Much additional information can be obtained from the Joint Nature Conservation Committee (JNCC) for **coastal zone** habitats, following a programme of research into the extent and quality of the major habitat groups, including:

- the Saltmarsh Survey of Great Britain (Burd 1989). A review of erosion and accretion processes on British saltmarshes has been commissioned by MAFF (Pye and French, 1993) and includes a database of all saltmarshes over 15ha in area;
- the National Sand Dune Survey (Radley, in press; Dargie, 1993, in press);
- the Shingle Survey of Great Britain (Randall et al 1990; Sneddon and Randall 1993);
- the Directory of Saline Lagoons and Lagoon–like Habitats (eg. Smith and Laffoley, 1992);

- the Estuaries Review (Davidson et al 1991).

For the North Sea Coast, the JNCC has compiled a directory of the conservation interests and an indication of the activities which have an effect on the coastline (Doody et al 1993); this directory is an important source of background information for coastal management.

Geological deposits which contain bulk materials can be identified from British Geological Survey maps and mineral assessment reports. Of particular relevance to erosion, deposition and flooding studies are the mineral assessment reports of on–shore areas (mostly undertaken by the BGS) and off–shore areas (all undertaken by the BGS) including:

- Marine sand and gravel resources off Great Yarmouth and Southwold. WB/88/9.

- Marine aggregate survey: south coast. WB/88/31
- Marine sand and gravel resources off the Isle of Wight and Beachy Head. WB/89/41
- Marine aggregate survey: cast coast. WB/90/17.
- Marine sand and gravel resources off the Humber. WB/92/1.

The locations of mineral assessment studies undertaken by the BGS are shown in Figure 5.3.

The Possible Effects of Development

Environmental concerns are a major factor in decision making. It is important, therefore, that policies, proposals and strategies should take account of the possible range of effects of development on the environment including, in this context, the operation of erosion, deposition and flooding processes. As was described in Chapter 2, significant effects may arise both at and around the **site**, and elsewhere within a catchment or coastal system. For example, the failure of a dam may result in severe consequences downstream; on the coast, the disruption of sediment transport caused by the construction of harbour breakwaters, etc., can lead to a deficit of sediment "downdrift" and adverse effects on coastal features such as dunes or beaches. Table 5.5 provides a range of examples of how development can have significant effects on erosion, deposition and flooding issues.

Although environmental assessment procedures are focused on the potential effects of specific proposals at a selected site, the **sediment budget and sensitivity** of the physical environment should be considered at a strategic level. In this way decision makers can determine, in broad terms, which types of development might be preferred or opposed in particular areas and the measures that may need to be taken to limit the effects. At this level, the key considerations will include:

(i)   the **linkages** between landforms within a catchment or coastal system. An important mechanism of linkage is through the transport of sediment from source areas to sinks where it is deposited in new landforms or as unconsolidated spreads on the sea or river beds (i.e. the sediment

budget). The sediment supplied from source areas can:

- affect the quality of water that is carried in a channel, influencing the rate of channel migration;

- affect the quantity of water that can be carried in a channel influencing the flood risk or the storage capacity of a reservoir;

- protect slopes or coastal cliffs from further erosion;

- protect low lying coastal areas from flooding, through the build up of sand dunes, mudflats, beaches and saltmarshes.

(ii)   the **sensitivity** of landforms to change. Individual landforms can be very responsive to changes in erosion or deposition patterns elsewhere in the landscape. For example, some river channels can readily adjust to changes in sediment supply by meander migration or the development of a braided network. However, the **sensitivity** to change is not constant; it can vary considerably between landforms and systems. On the coast, for example, sensitive systems can be recognised on the basis of the strength of the littoral drift and the reliance of the individual landforms on a continued sediment supply.

Within any system, however, there will be stable elements, such as rocky headlands, and unstable elements such as relict beaches. A stable system is one in which the controlling resistances (i.e. strength of materials, equilibrium form, structural resistance) are such as to either prevent an event from having any effect or to be so arranged as to restore the system to its original state. For example, although wind and wave energy is greatest on the north and west coasts this does not correspond with rapid erosion and sediment transport because of the resistant rocks along the shoreline.

Figure 5.3 Areas covered by bulk mineral assessment reports (BGS, 1994).

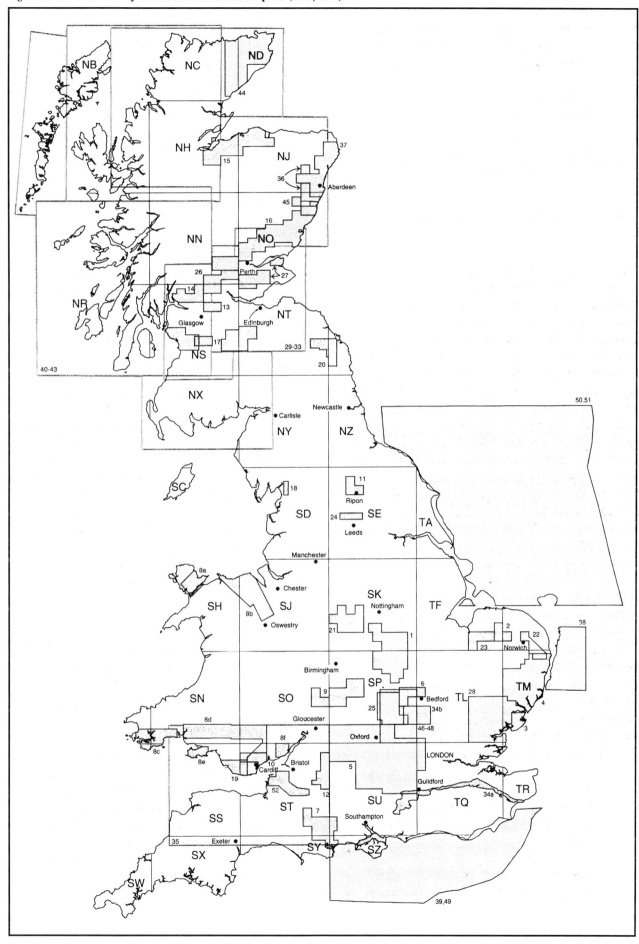

Table 5.5 A selection of the possible effects of development on aspects of erosion, deposition and flooding issues.

SYSTEM	POTENTIAL EFFECTS
Hillslope	• increase in runoff • increased infiltration through soakways etc. and consequential effects on slope stability. • creation of urban flood problems when rainfall intensity exceeds storm-drain capacity. • modifications to channel form can lead to local intensification of flood problems. • bridges and culverts can be sites of temporary dams in flood events.
River	• development downstream of a reservoir can in some cases result in the reservoir owner having to improve dam safety. • increase in peak flow and reductions in sediment supply can lead to channel change, especially increase in width and/or depth. • effects on river corridor habitats and geomorphological features. • reduction in floodplain storage and infiltration. • land reclamation can lead to inter-tidal squeeze. • flood embankments can cause loss of flood wave attenuation and lead to more severe flooding downstream (floodplain areas) or upstream (estuaries). • dredging may lead to an increase in sedimentation.
Coasts	• coastal defences, breakwaters etc. may disrupt sediment supply to beaches, dunes, mudflats etc. • mineral extraction from beaches and dunes or clay digging in estuaries may lead to decline in natural coastal defences and habitats. • land reclamation or coastal defences can lead to loss of inter-tidal habitats. • uncontrolled surface water drainage, leaking water pipes or sewer systems can lead to increased landsliding. • excavation can lead to a reduction in slope stability. • recreational pressures can lead to destabilisation of dune systems.

Background information about the nature and sensitivity of physical systems can be difficult to obtain, especially because most studies of the behaviour of such systems tend to be focused towards individual sites rather than broad areas. However, in forthcoming years these problems may be eased by the programmes of broad based studies by the NRA (catchment management plans) and coastal defence groups (shoreline management plans) in England and Wales.

The Need for Monitoring

Local planning authorities are required to keep under review the principal physical characteristics of their area (under the Town and Country Planning Act 1990); other authorities will also need to monitor the operation of the processes on their interests or to evaluate the effects of global warming and sea level rise. Coastal defence groups, for example, have been advised by MAFF/Welsh Office that shoreline management plans will need to be reviewed at regular intervals because they can reflect only **current levels** of knowledge of physical processes and technical responses. Indeed, much of our understanding of the operation of the

processes is based on an awareness of past events. Predictions of climate changes over the next few decades indicate that judging the future pattern of events from what happened in the past may be unwise. As a result, the general background erosion, deposition and flooding information may need to be regularly reviewed and updated, by means of:

• maintaining records of damaging events;

• regular review of ground conditions (e.g. channel positions, rate of river and coastal erosion, state of natural coastal defences, etc.);

• monitoring any changes in the likelihood of potentially damaging events arising as a result of climate change, i.e. change in risk.

Summary

General assessments of the erosion, deposition and flood character of the area covered by an authority and beyond can provide guidance to decision makers on:

- where erosion, deposition and flooding issues need to be considered;

- the nature of the hazards in the areas and the general level of risk to existing land uses and development;

- whether potentially damaging events can be expected during the normal lifetime of particular types of development;

- the types of uses or development that, in general terms, may be best suited to those areas;

- whether development or land use can have adverse effects on other interests in the river or coastal environments.

At this broad level of consideration, investigations should, essentially, be **desk studies** comprised of compilations and analyses of readily available information supplemented by inspection of historical maps or aerial photographs, where necessary. It will not be feasible to collect and collate all existing data; to do so may be counter-productive as the need is for a broad brush appreciation of the physical character of an area and not detailed information about well researched sites.

A range of basic sources of information exist that will usually provide the most relevant data on the operation of erosion, deposition and flooding processes. These have been described in this Chapter. However, lack of awareness of the potential for damaging events in an area is, however, often a significant problem, especially where the threat is associated with extremely rare events or failure of man–made structures. It is important, therefore, that the first step in any assessment should be to consider whether any of the characteristic problems associated with specific terrain units are relevant in a particular area (Table 4.1). Once the potential for natural hazards has been recognised, defining the extent of vulnerable areas can, in most cases, be readily achieved from available information sources. Specialist advise should, however, be sought in many circumstances from the variety of bodies and authorities involved with the collection and analysis of erosion, deposition and flooding information (Table 4.3). Decision makers will also need to be aware as to whether elements of the natural environment are

dependant on the continued operation of physical processes, as development or construction of defence works may have significant adverse effects on the value of these features.

Chapter 5: References

British Geological Survey, 1994. Catalogue of Maps. BGS Keyworth.

Burd F 1989. The Saltmarsh survey of Great Britain: An inventory of British Saltmarshes. NCC Research and Survey in Nature Conservation No.17.

Dargie T C D 1993. Sand dune vegetation survey of Great Britain. Part 2: Scotland. JNCC.

Dargie T C D, in press. Sand dune vegetation survey of Great Britain. Part 3: Wales, JNCC.

Davidson N C, d'A Laffoley D, Doody J P, Way L S, Gordon J, Drake C M, Pienkowski M W, Mitchell R and Duff K L 1991. Nature conservation and estuaries in Great Britain. NCC.

Doody J P, Johnston C and Smith B, 1993. Directory of the North Sea Coastal Margin. JNCC.

Englefield H.C. 1816. A description of the principal picturesque beauties, antiquities and geological phenomena of the Isle of Wight. London.

Geomorphological Services Limited 1986/87. Review of Landsliding in Great Britain. Reports to DoE.

Jones D K C and Lee E M, 1994. Landsliding in Great Britain. HMSO.

Lee E.M. and Moore R. 1991. Coastal landslip potential assessment Ventnor, Isle of Wight. Report to DoE.

Lee E.M., Moore R., Burt N. and Brunsden D. 1991. Strategies for managing the landslide complex of Ventnor, Isle of Wight. In R.J. Chandler (ed) Slope stability engineering: development and applications, 189–194. Thomas Telford.

Matthews W. 1950. British Diaries. An annotated bibliography of British Diaries (1442–1942). University of California Press.

Ministry of Agriculture, Fisheries and Food 1994. Coast Protection Survey. MAFF.

Norton J.E. 1950. Guide to the national and provincial directories of England and Wales, excluding London, published before 1856. London.

Pickles L. and Woolhouse C. 1994. Catchment management plans: a strategic view of flood defence. In MAFF Conference of River and Coastal Engineers.

Potter H.R. 1978. The use of historical records for the augmentation of hydrological data. Institute of Hydrology Report No. 46.

Pye K and French P W 1993. Erosion and accretion processes on British Saltmarshes, volumes 1–5. Cambridge Environmental Research Consultants Ltd.

Radley G P, in press. Sand dune vegetation survey of Great Britain. Part 1: England. JNCC.

Randall R E, Sneddon P and Doody J P, 1990. Coastal shingle in Great Britain, a preliminary review. NCC Contract Survey No.85.

Smith B P and d'A Laffoley D 1992. Saline lagoons and lagoon –like habitats in England. English Nature.

Sneddon P and Randall R E 1993. Coastal vegetated shingle structures of Great Britain: Main Report. JNCC. Appendix 1 – Wales; Appendix 2 – Scotland; Appendix 3 – England.

White W. 1881. History, gazetteer and directory of Lincolnshire. Sheffield.

6 Site Reconnaissance Surveys

Introduction

The selection of sites for development should involve an assessment of the relative risks imposed by erosion, deposition and flooding processes. Such considerations are relevant to:

- local planning authorities who need to identify suitable sites for development or redevelopment in Local Plans (or UDP Part IIs) i.e. a **site review**;

- developers who need to have advance warning of the potential difficulties associated with particular sites (i.e. **preliminary assessment**).

Both planners and developers have an interest in the selection of sites for development and thus may need to undertake appropriate site studies. **Site review** should be carried out by planners to determine whether the proposed sites are in vulnerable locations and, if so, the likelihood of potentially damaging events. The need for new defences or improvements to existing defences should be identified, assessing whether such works are likely to be cost effective or environmentally acceptable. Such considerations are not restricted to developers; the nature of the "plan led" planning system dictates that development plan proposals should take full account of potential problems as failure to do so could lead to development in unsuitable areas. A **preliminary assessment** may also provide advance warning to developers of the potential difficulties associated with particular sites and an indication of the cost implications.

Much of the necessary information may be collated from previously collected data, especially the results of **general assessments** or **shoreline management plans** (Chapter 5). These sources may provide ready access to information that may be relevant to some or all of the shortlisted sites.

However, as these are essentially broad–brush studies they will need to be supplemented by specific information about conditions at the individual sites. To establish the treats from erosion, deposition and flooding (i.e relative **risks**) and possible environmental effects of development (i.e **sediment budget** and **sensitivity**).

In areas where erosion, deposition and flooding are likely to constrain the development of a site, decision makers will need to have access to information on whether the site can be safely and cost effectively developed without adverse effects on other interests. They will need to be aware of the nature and extent of the problems that may occur, the likelihood of damaging events, the possible risks to property, infrastructure and services in the vulnerable areas and the measures that can be adopted to minimise these risks. This requires an assessment of:

- the boundaries of the land at risk from flooding or instability;

- the scale and history of potentially damaging events;

- the likelihood of future damaging events and the possible impact on the sites;

- the nature, scale and potential costs of any defence works necessary to reduce the risks to an acceptable level;

- the possible effects of the development and associated defence works on the level of risk to neighbouring areas or environmental interests.

Desk Study

The first stage in any site reconnaissance should be a **desk study** of available records, maps and reports in order to:

- gain as much background information about risks, sediment budgets and sensitivity as possible;

- reduce the amount of detailed ground investigation that may be required, by allowing resources to be targeted towards key issues;

- allow efficient planning of the ground inspection;

- allow cost-effective design of subsequent investigations.

The desk study should concentrate on a range of factors relating to the ground conditions at the shortlisted sites (Table 6.1). Similar principles apply to those described for a general assessment (Chapter 5) although there will need to be a greater level of detail on the specific sites under consideration. The study should be focused on the conditions at the relevant sites, although it will also need to consider the broader setting to establish whether processes operating on adjacent slopes, or elsewhere in the catchment or coastal system could affect the site or vice versa.

The desk study should be accompanied by some form of **ground inspection** to confirm or determine the nature of the conditions at the different sites. This inspection should complement the desk study and allow **preliminary assessments** of **risk** and the possible **environmental effects** to be made. From this a judgement can then be made about whether the site or sites require defence works and, if so, the likely scale and costs of those works.

Aerial Photograph Interpretation (API)

Since the 1950's aerial photographs have been used extensively in site investigations in Great Britain. Aerial photography provides an exact and complete record of the ground surface at a given moment in time and hence represents the most efficient means of recording natural and anthropogenic features of the terrain. The underlying philosophy behind API is the presentation of landscape features in a convenient format and the concept that stereoscopic vision, in the case of a river channel or eroding cliff, allows the creation of a tree-dimensional image that represents exactly the details of the form of the ground surface, within the constraints of scale and photograph quality. Common scales of air photography range from 1:5,000 and larger for detailed studies to smaller than 1:25,000. Photographs are normally taken in succession as 'sorties', until the whole area under investigation is recorded. A 60% + overlap between successive photographs allows stereoscopic vision when viewed through a stereoscope.

Information obtainable from aerial photograph interpretation includes the location of drainage channels and associated drainage patterns, variability in underlying geology and soil types, slope soil types, slope inclination data (if stereoscopic vision is used), and areas prone to flooding or potentially vulnerable to erosion or landsliding. Clearly then, air photographs can provide information for relationships between bedrock, soil, drainage and slope stability, qualitative slope and drainage data, land use and existing and planned engineering structures. Preliminary API can not only provide information directly relevant to the investigation but also may significantly increase the efficiency with which the site investigation is planned. The use of air photographs in the preliminary design of site investigation for all engineering works has become well integrated into codes of practice in Great Britain and is briefly described in the Site Investigation Code of Practice BS5930 (BSI, 1981).

The principal advantages of API for erosion, deposition and flooding studies include allowing:

- boundaries of floodplains, coastal lowlands and landslide complexes to be quickly determined;

- appreciation of surrounding slope conditions;

- rapid measurement of land loss if aerial photographs of different dates are available.

Ground Inspection

Irrespective of the availability of background information, there will usually be a need for ground inspection to confirm desk study results.

Table 6.1 General information required for a desk study (after BSI, 1981).

1.	General Land Survey
	• Ordnance Survey Maps and historical maps
	• British Geological Survey Maps and Memoirs
	• Soil and Land Research Centre Maps and Memoirs (England and Wales)
	• Macaulay Land Use Research Institute Maps and Memoirs (Scotland)
	• Admiralty charts and hydrographic publications, including tide tables
	• National Landslide Databank (Rendel Geotechnics)
	• Flood Studies Report (NERC, 1975)
	• Macro Review of the Coastline of England and Wales (HR Wallingford various dates)
	• UK Digital Marine Atlas (Institute of Oceanographic Sciences)
	• Catchment Management Plans (NRA)
	• Shoreline Management Plans (Coastal Defence Groups)
2.	Meteorological Information (Meterological Office)
	• daily rainfall records, including continous chart records
	• monthly weather reports
	• British Rainfall – monthly and annual rainfall totals for around 6000 stations
3.	Hydrological Information
	• Surface Water Yearbook – published annually for selected gauging stations (now Hydrological Data UK)
	• The Surface Water Archive (Institute of Hydrology, Wallingford)
	• The Flood Event Archive (Institute of Hydrology, Wallingford)
4.	Remote Sensing
	• satellite imagery (NRSC Farnborough)
	• aerial photographs (Ordnance Survey and private firms)
5.	Drainage and Sewerage
	• local authority and water company records
6.	Coastal Information
	• sea level data (Proudman Oceanographic Laboratory)
	• currents (Proudman Oceanographic Laboratory)
	• waves and tides (Proudman Oceanographic Laboratory)

Surface mapping techniques such as geomorphological mapping can be used to establish the nature and extent of possible hazards in an area and to determine the degree of threat that these hazards may pose to existing property and future developments. Although geomorphological mapping has been developed for the investigation of slope instability problems (e.g. Figure 6.1) it can be adapted to provide useful site information for other purposes (Table 6.2). In such cases, attention should be focused on identifying signs of potential instability (e.g. for sand dunes) the extent of floodplains and former floods (e.g. for flash floods), possible sites of channel instability, bank overtopping or breaching of flood defences (e.g. for lowland flooding). Geomorphological surveys of catchments and coastlines can help pinpoint sources of sediment and historical causes of contemporary change.

It is important to recognise, however, that there are significant limitations in surface mapping techniques, not least because they can only identify static features and cannot be relied upon to define the dynamic nature of the environments under consideration. For this reason, it is important that surface mapping is accompanied by an assessment of the **potential for change**. This can be achieved by examining the morphology of the landforms; some landform types are more prone to change than others. An historical perspective can also be very useful. Here, a general indication of the nature and scale of surface changes that have occurred to features such as river channels or natural coastal defences can be determined by comparing of historical sequences of maps, postcards and aerial photographs, (Table 6.3).

Figure 6.1 Surface mapping techniques: morphological mapping (top; after Cooke and Doornkamp, 1990); Geomorphological mapping (bottom; after Lee and Moore, 1989).

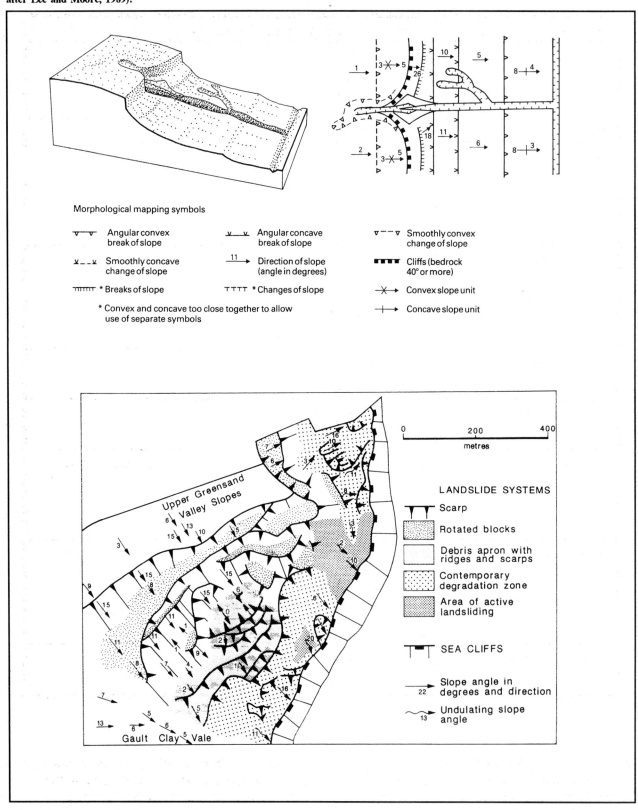

Geomorphological mapping can be regarded as an essential tool for rapid site inspection, portraying surface form together with an **interpretation** of the origin of particular forms and features i.e related to geological conditions, surface processes or human activity. The technique was used, for example, as part of an investigation to establish the cause of the 1987–88 landsliding at Luccombe, Isle of Wight which had resulted in a number of houses being damaged beyond repair (Lee & Moore 1989). By

Table 6.2 Geomorphological mapping; defining surface forms and signs of active processes.

PROBLEM	GEOMORPHOLOGICAL FEATURES FOR MAPPING
Hillslope	• gulleys and rills • debris fans • dry valleys
Wind Erosion	• shelter belts
Flash Flooding	• former strand lines or debris spreads • sites for potential debris dams • upstream boulders etc. • channel geometry
Channel Instability	• actively eroding or accreting reaches • sources of sediment • former channels • sites for potential meander cut offs
Lowland Flooding	• extent of floodplain • sites for potential overtopping or breaching
Estuary Sedimentation	• extent of mudflats and saltmarshes • signs of active erosion or accretion
Coastal Flooding	• nature and form of "natural defences", e.g. mudflats, salt marshes, beaches, shingle banks or sand dunes • signs of active erosion or accretion • extent of low lying areas
Coastal Erosion	• nature and extent of coastal landslides • sources of sediment • signs of activity, e.g. tension cracks, fresh scars etc. • drainage features
Wind Erosion in Dunes	• nature and extent of dune system • signs of active instability, e.g. blow outs etc.

carefully mapping the limits of the recent movements, together with the surface form of the surrounding area, it was possible to demonstrate that the village had been built on an ancient landslide and that the recent movements were a reactivation of parts of this feature (Figure 6.1).

It must be recognised, however, that such qualitative assessments can be fallible, especially for landslide assessments. Slopes may be so remodelled by human activity or natural processes that no surface indicators exist to indicate the presence of failed masses and buried shear surfaces. Conversely, prominent tension cracks on highly irregular ground may be the result of superficial slippage on generally stable landforms produced by quite different processes. Thus, irrespective of the interpretive skills brought to bear, it is generally necessary to confirm visual diagnoses by subsurface investigation (Chapter 7).

Table 6.3 Geomorphological mapping: monitoring change from old aerial photographs, postcards and Ordnance Survey maps.

PROBLEM	FEATURES TO BE MAPPED
Hillslope Erosion	• changes in the pattern and density of gulleys and rills • growth of debris fans etc.
Flash Flooding	• change in positions of major boulders etc. • former strand lines • changes in channel form
Channel Instability	• changes in the channel pattern • erosion or accretion of riverbanks, channel bars etc.
Coastal Flooding	• changes in the nature and extent of natural defences, especially the seaward and landward edges
Coastal Erosion	• changes in the position of cliff top • scale of previous landslide feature • the pattern of development of landslide units • drainage features
Wind Erosion in Dunes	• changes in the position of the seaward and landward edges of dune systems • evidence of previous instability, e.g. blow outs • changes in the pattern of vegetation

Preliminary Assessment of Risk

Although the approach to hazard and risk assessment will vary with the nature of the problems that could be encountered at the site or sites (Table 6.4), there are a number of factors that are common to most investigations:

• what sizes event can be expected at a site;

• the casual factors associated with different events;

• how frequently can the event triggering conditions be expected;

• what area would be affected by a particular event;

• what property, infrastructure and services are located in the vulnerable areas.

Flood Hazard Assessment

From a hydrological perspective, **flood hazard assessment** involves the statistical assessment of the probability of a discharge of a given size occurring over a particular time period (the return period; Chapter 4). Such **magnitude/frequency** studies are central to many aspects of flood research and the design of flood defences. Other studies are required when engineers need to know about the duration and volume of flooding, as for the design of storage reservoirs.

The **Flood Studies Report** (NERC 1975) is a major source of information for both flood records and methods of statistical analysis. The study was sponsored by the Natural Environment Research Council (NERC), and was undertaken at the Institute of Hydrology in the early 1970s. Records were collected for all parts of the country and the result is a major source of flood information. The hydrometeorlogical studies were based on analyses of records from more than 600 daily rainfall stations with an average of 60 years of records supplemented by daily and monthly records from 6000 stations for the period 1961–70. All available

Table 6.4 Approaches to the assessment of different erosion, deposition and flooding hazards.

PROBLEM	APPROACH TO HAZARD ASSESSMENT
Hillslope Erosion	• what rainfall conditions could initiate significant erosion under particular land management practices • how often do these rainfall conditions occur
Wind Erosion	• what wind speeds could initiate significant erosion under particular land management practices • how often do these wind speeds occur
Flash Flooding	• what surface runoff can be expected to be generated by particular rainfall events in a particular sub–catchment • is the existing channel network (natural or artificial) capable of carrying the runoff from particular events • what size material is likely to be carried by the flood event; is there a ready supply of such material • is there a risk of dam failure
Channel Instability	• what forms of channel change can be expected under extreme flow conditions • what is the average annual bank erosion rate • what size events are involved in bank erosion; how often do they occur
Lowland Flooding	• what discharges can be expected under particular rainfall or snowmelt conditions • will channel capacity be exceeded by particular discharges • what area will be affected by a particular flood event and to what depth • are there sites for potential channel overtopping, e.g. undersized culverts etc.
Estuary Sedimentation	• what is the average annual sedimentation rate • what is the range of annual sedimentation rates • is the sedimentation rate linked to river discharge, land use change in the catchment or dredging operations
Coastal flooding	• what combined tidal, storm surge and river flow conditions can be expected over a particular time period • are the natural defences capable of withstanding extreme storm surge conditions • are there sites of potential breaches in existing defences • what area will be affected by a particular flood event and to what depth
Coastal Cliff Erosion	• what size events are involved in coastal erosion along the entire length of the outcrop of a particular method • what is the average annual cliff erosion rate along the outcrop • how often do significant events occur at a particular location along the outcrop • what rainfall and tidal conditions can be expected to initiate coastal instability
Wind erosion of Sand Dunes	• what wind speeds could initiate significant erosion under particular management practices • how often do these wind speeds occur

flood flow records were also collected and information from 533 sites was used which, in total, yielded some 6000 station–years of records. The Report makes available most of the flood flow information for Britain and can be used to obtain flow data for most parts of the country. Mean daily and peak flow data are retained by the NRA and private Water Companies but the main source of flood data remains the Institute of Hydrology, Wallingford, with their constantly updated **Flood Data Archive**.

Two approaches are proposed by the Flood Studies Report for magnitude/frequency estimation:

(i)   **flood frequency analysis** based on a statistical approach to peak flood flows based on catchment characteristics, modified by historical data, where available;

(ii)   **unit hydrograph analysis**, a deterministic approach involving synthesis of the flood corresponding to a design storm.

Prior to the Flood Studies Report, hydrologists used **rational methods** for predicting maximum flows. These methods are based on attempts to relate maximum discharge to catchment and precipitation characteristics. For example, probably the most widely used formula in Great Britain has been the method outlined in the Institution of Civil Engineers' 1933 Committee Report which had been written to aid the execution of the **Reservoirs (Safety Provisions) Act 1930**. For upland catchments the formula is:

Normal Maximum Flood discharge = $1000 \times$ Catchment Area$^{0.6}$

The rational methods can still provide a useful check for results derived from other methods of flood analyses. Indeed, many hydrologists use more than one method of analysis and compare them to provide the best solution.

The Flood Study also identified that the relationship between mean annual flood and the flood of a particular return period varied across the country; in lowland areas, extreme events such as a 100 year flood may be 3 to 4 times the mean annual flood (Figure 6.2). In upland areas a similar return period event may be only twice the size of the annual flood. This pattern is largely a function of catchment character and has major implications for the size of defence works needed to protect against floods of a give return period.

With several years of flood recording available since the publication of the Flood Studies Report in 1975, and much supplementary research (e.g. ICE, 1981), a successor publication – the **Flood Estimation Handbook** – is in preparation. Publication will be in five volumes to be released between 1995 and 1998; specific additional research needed to meet these targets will be funded by MAFF and the NRA.

Assessment of Coastal Floods

The magnitude and frequency of extreme events can be predicted from the historical record of storm surges, using statistical methods for assessing the heights of high tides and the meteorological conditions likely to give rise to surges (Table 6.5; Figure 6.3). However, the longest continuous sets of records cover only around 100 years; combined river flow and tidal records for analysis of surges in estuaries are much shorter, often dating from around the mid 1950s. Notable problems arise when considering the combined effects of storm surges and tides and waves. **Joint probability assessments** ideally need to be based on long period records of wave height and still water level. Specialist advice can be obtained, for example, from the **Proudman Oceanographic Laboratory**.

The Storm Tide Warning Service was established after the 1953 east coast floods to link hydrographic and meteorological information and produce accurate **predictions** of coastal flood levels during a storm event. This system is described in Chapter 11 and is based on the use of:

(i)   **empirical equations** derived from a regression analysis of data sets compiled from a large number of previous surges. They provide site specific forecasts of high water residuals and are only available for locations on the East Coast. These equations use the observed surge at more northerly ports as a starting point, and have pressure and wind term to calculate the modulation as the surge travels south, (Pratt, 1993);

(ii)   a **numerical surge model** first became operational in 1978/79 with periodic upgrades being made since; since 1991/92 it has been based on a 12km grid. There are a number of ways forward with surge forecasting, including combined wave and surge models, the assimilation of observed

Figure 6.2 The relationship between return period and the size of flood event, as related to the mean annual flood, for 10 regions in Great Britain (after NERC, 1975).

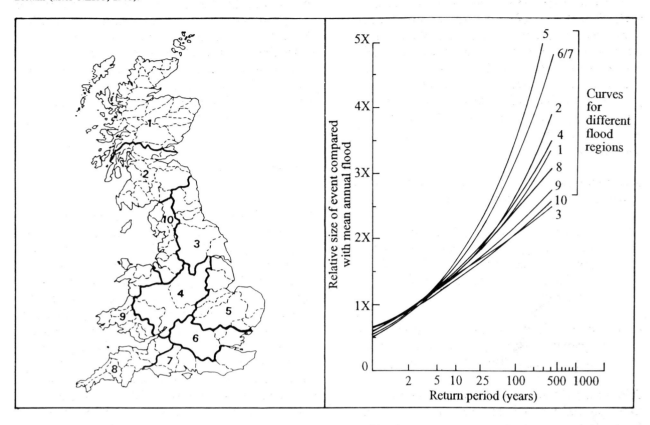

Figure 6.3 Methods employed to forecast coastal floods (after Pugh, 1987).

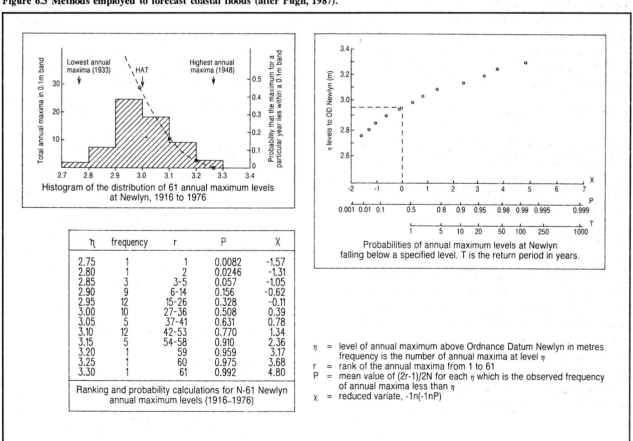

Histogram of the distribution of 61 annual maximum levels at Newlyn, 1916 to 1976

Probabilities of annual maximum levels at Newlyn falling below a specified level. T is the return period in years.

η	frequency	r	P	χ
2.75	1	1	0.0082	-1.57
2.80	1	2	0.0246	-1.31
2.85	3	3-5	0.057	-1.05
2.90	9	6-14	0.156	-0.62
2.95	12	15-26	0.328	-0.11
3.00	10	27-36	0.508	0.39
3.05	5	37-41	0.631	0.78
3.10	12	42-53	0.770	1.34
3.15	5	54-58	0.910	2.36
3.20	1	59	0.959	3.17
3.25	1	60	0.975	3.68
3.30	1	61	0.992	4.80

Ranking and probability calculations for N-61 Newlyn annual maximum levels (1916–1976)

η = level of annual maximum above Ordnance Datum Newlyn in metres
frequency is the number of annual maxima at level η
r = rank of the annual maxima from 1 to 61
P = mean value of (2r-1)/2N for each η which is the observed frequency of annual maxima less than η
χ = reduced variate, $-\ln(-\ln P)$

75

Table 6.5 Sources of tide, current and wave data.

Sea Level Data	•	the UK national tide gauge network, established after the 1953 floods
	•	records are processed at Proudman Oceanographic Laboratory (POL) Bidston to provide hourly series of observed and surge data
Current Data	•	no permanent current meter stations, data collection usually carried out during specific studies
	•	POL maintains a databank of current data
Wave Data	•	routine wave data acquisition by Light Vessels ("Channel", "Sevenstones") and coastal sites
	•	inventory of available instrumentally – recorded wave data published by POL

surge levels into models and finer mesh local area surge models.

Coastal Landslide Hazard and Recession Potential

Assessment of landslide hazard and recession potential involves the evaluation of the key factors which influence the pattern and character of landslide activity:

- geological materials;
- topography;
- climate;
- hydrogeology;
- wave and tide climate;

The variable interaction amongst specific aspects of these factors helps to determine the precise siting of instability, the timing of failure and to dictate the condition of displacement in terms of the form, rate and duration of movement. However, there are a broad range of causal factors that may be important; some trigger instability through short term fluctuations, while others act in an insidious manner to slowly reduce the resistance of a cliff to failure.

The great variety of landslide forms and causes means that landslide hazard assessment techniques have to be adapted to the local conditions. However, a number of principles are likely to apply in most cases:

- establish the type and size of failure that could occur at the selected sites. This should involve identifying the nature and extent of past and present failures along the entire outcrop of the materials that underlie the selected sites;

- establish a historical record of events, their size and the impact on property, services and infrastructure;

- establish a relationship between historical events and rainfall patterns or storm tide levels. This will enable a broad assessment to be made of the likelihood of an event occurring somewhere on the outcrop of the site materials as a result of a particular rainfall period or wave height;

- establish what areas could be affected by further landsliding and over what time period.

These principles are well illustrated by the DoE – commissioned assessment of Coastal Landslip Potential for Ventnor, Isle of Wight (Lee & Moore, 1991). Ventnor is an unusual situation in that the whole town lies within an ancient landslide complex. The problems are related to the control of the nature of development in those parts of the town which have been shown to be particularly susceptible to ground movement. The hazards faced by the local community in Ventnor have been defined in terms of an understanding of **contemporary ground behaviour** within an extensive belt of landslipped ground.

The approach developed for assessing the hazards at Ventnor involved a thorough review of available records, reports and documents relating to instability, followed by a detailed field investigation comprising geomorphological and geological mapping, photogrammetric analyses, a survey of damage caused by ground movement, a land use survey and a review of local building practice (Figure 6.4). The results of these

Figure 6.4 The work programme for the Ventnor study (after Lee and Moore, 1991).

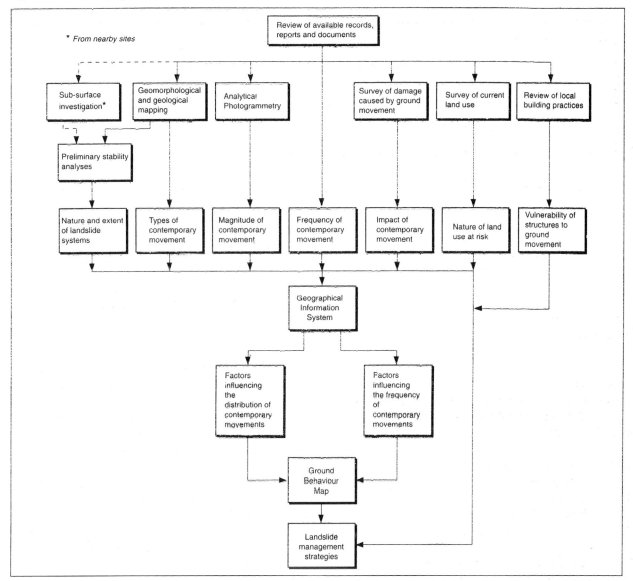

investigations provided an understanding of the nature and extent of the landslide system, together with the type, size and frequency of contemporary movements and their impact on the local community. This detailed understanding of ground behaviour was used in conjunction with knowledge of the vulnerability to movement of different types of construction and the spatial distribution of property at risk, to formulate a range of management strategies designed to reduce the impact of future movements.

Geomorphological mapping revealed the scale and complexity of the landslides; once the framework of landslide units had been established, it was possible to relate building damage and movement rates to units with known dimensions. A search through historical documents, local newspapers from 1855–1989, local authority records and published scientific research, revealed nearly 200

individual incidents of ground movement over the last two centuries. The landslide–related information were presented as a 1:2500 scale map of Ground Behaviour which summarised the nature, magnitude and frequency of contemporary processes and their impact on the local community. The ground behaviour map demonstrated that the problems resulting from ground movement vary from place to place according to the geomorphological setting.

This formed the basis for **landslide management strategies** that can be applied within the context of a zoning framework that reflects the variations in stability rather than a blanket approach to the problem. In support of the management strategy a 1:2500 scale Planning Guidance map was produced which related categories of ground behaviour to forward planning and development control (Figure 6.5, Table 6.6). The map indicated that different

Figure 6.5 Summary planning guidance map, Ventnor (after Lee and Moore, 1991).

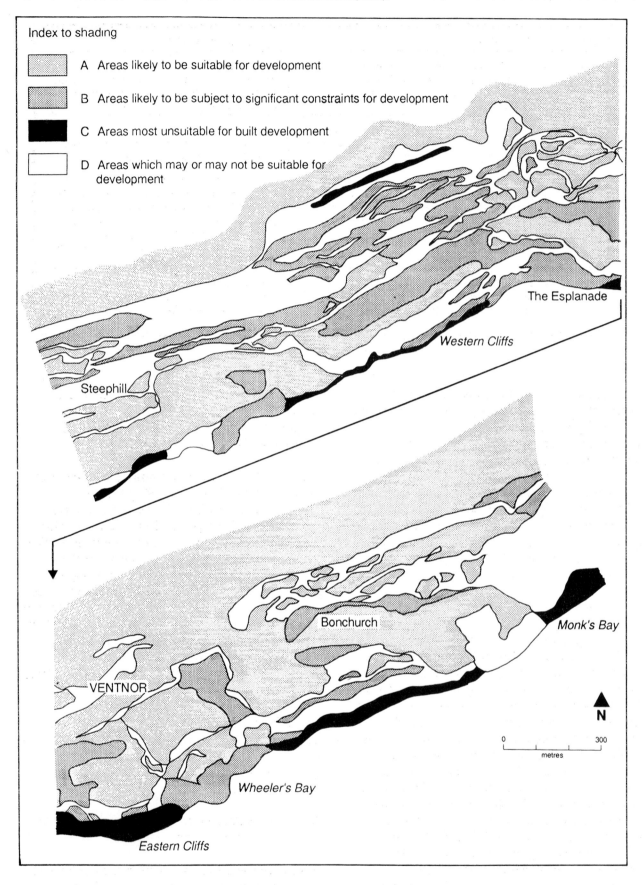

Index to shading

A Areas likely to be suitable for development

B Areas likely to be subject to significant constraints for development

C Areas most unsuitable for built development

D Areas which may or may not be suitable for development

The Esplanade

Western Cliffs

Steephill

Bonchurch

Monk's Bay

VENTNOR

N

0   300
metres

Wheeler's Bay

Eastern Cliffs

Table 6.6 Planning guidance categories for management of landslide problems in Ventnor, Isle of Wight (after Lee and Moore, 1991).

CATEGORY	DEVELOPMENT PLAN	DEVELOPMENT CONTROL
A	Areas likely to be suitable for development. Contemporary ground behaviour does not impose significant constraints on Local Plan Development proposals.	Results of a desk study and walkover survey should be presented with all planning applications. Detailed site investigations may be needed prior to planning decision if recommended by the preliminary survey.
B	Areas likely to be subject to significant constraints on development. Local Plan development proposals should identify and take account of the ground behaviour constraints.	A desk study and walkover survey will normally need to be followed by a site investigation or geotechnical appraisal prior to lodging a planning application.
C	Area most unsuitable for built development. Local Plan development proposals subject to major constraints.	Should development be considered it will need to be preceeded by a detailed site investigation geotechnical appraisal and/or monitoring prior to any planning applications. It is likely that many planning applications in these areas may have to be refused on the basis of ground instability.
D	Areas which may or may not be suitable for development but investigations and monitoring may be required before any Local Plan proposals are made.	Areas need to be investigated and monitored to determine ground behaviour. Development should be avoided unless adequate evidence of stability is presented.

areas of the landslide complex need to be treated in different ways for both policy formulation and the review of planning applications. Areas were recognised which are likely to be suitable for development, along with areas which are either subject to significant constraints or mostly unsuitable. This example demonstrates the relevance of landslide information to planning, it also provides a model for addressing other risks within the context of the planning and development process, most notably flooding.

Risk Assessment

At the site selection stage it is often inappropriate to evaluate risk in absolute terms because of the difficulties in confidently assigning values for the hazard, value of property at risk and vulnerability. In many cases it is more useful to assess the **relative risk** to each of the shortlisted sites from particular hazards, based on both factual data and subjective appraisal. The value of relative risk assessment is that it can quickly enable sites to be compared or allow decisions to be made about where limited resources and finances should be directed. In the latter case it is important to determine how long the situation can be left untreated before action becomes essential and whether by delaying, the options for treatment become more limited. For example, cliff treatment measures involving laying a slope back to a more

stable angle require ample space in front of the property at risk; unchecked erosion will progressively reduce the available space until laying back is no longer a viable option.

An example of the use of a relative risk approach is provided by Clark et al (1993) for the management of weak sandstone cliffs at Shanklin, Isle of Wight. Here potential cliff face and talus slope failures were evaluated by ground inspection of a series of cliff sections, enabling a systematic assessment to be made of the hazard along the entire cliffline. The risk posed by the failures was established by assigning estimated hazard values to each of the cliff sections. These sections were then categorised into a rating based on qualitative assessment of commercial value, cost of reinstatement and the potential for injury or fatality if a large failure occurred.

The vulnerability of the land use elements was established by considering:

(i)   the distance of the elements from the hazard;

(ii)   the vulnerability to damage, as related to construction type.

Having assigned figures to both hazard, estimated value and vulnerability, the relative risk was calculated as the product of the three figures, i.e.:

Risk = Hazard x Value x Vulnerability

Preliminary Assessment of Environmental Considerations

Although formal environmental assessment is only required with major development projects included within the 2 schedules attached to the Town and County Planning (Assessment of Environmental Effects) Regulations 1988, it is in the interests of the developer that the environmental implications are assessed at an early stage of preparing a development proposal. This will generally involve identification of:

● the potential effects of the proposed development on the operation of physical processes including sediment transport;
● those landforms, habitats or features that are sensitive to changes in the rate of physical processes;
● the possible consequences of these changes on the natural environment;
● measures that can be employed to minimise the adverse effects on the environment.

In many instances it may be appropriate to carry our a preliminary environmental study to establish the key concerns that are likely to be generated by a proposed development, and if necessary, set the objectives of any subsequent environmental assessment. This type of preliminary study (often known as a scoping study) should involve consultation with key interest groups and authorities, and the qualitative appraisal of the possible environmental impacts. Specific environmental impact matrices can be developed to identify the more significant issues that could arise. "**Environmental receptors**" should be defined having regard for the characteristics of the development and the area which it covers; these may include: recreation, cultural heritage, flora and fauna, social and economic, landscape, etc (Figure 6.6). For major developments or flood and coastal defence schemes, it would generally be necessary to assess the potential impacts of a range of project options including, if appropriate, the "do nothing" option.

Summary

Reconnaissance surveys of sites for development should be undertaken by both local planning authorities and developers. They can help ensure that:

● land is not allocated for development in unsuitable locations (**site review**);

● new defence works or improvements to existing defences can be justified on economic and environmental grounds before the decision is made to proceed with the development (**preliminary assessment**).

Such investigations should involve a combination of the desk study approach described in Chapter 5, aerial photograph interpretation, ground inspection, preliminary assessments of risk and a scoping study to establish the possible environmental effects. Ground inspection should involve both a surface mapping exercise and an evaluation of the nature and scale of the changes that have occurred in dynamic environments over historical times.

Preliminary hazard and risk assessments may involve statistical or deterministic methods, but should address a number of key factors:

● what size events can be expected at a site;

● the casual factors associated with different events;

● how frequently can the event triggering conditions be expected;

● what area would be affected by a particular event;

● what property infrastructure and services are located in vulnerable areas.

Relative risk assessment of sites may be more appropriate than detailed analyses of the absolute risk. In this way, sites can be compared in terms of the suitability for development from both economic and environmental perspectives.
A range of basic sources of information and assessment techniques are available for use in Britain; the Flood Studies Report (NERC, 1975), and its subsequent revisions (e.g. ICE, 1981), is an essential tool in the assessment of flood hazard and can be used throughout Britain. Landslide hazard

The resulting risk values were rationalised into six relative risk categories to identify those areas requiring priority stabilisation works.

Figure 6.6 Sample matrix for an environmental scoping study for identifying potential significant impacts.

ENVIRONMENTAL RECEPTORS — Wall and Embankment Improvements: North Muirton Flood Embankment, Option A (A4)

PROJECT ACTIVITIES	Recreation	Cultural Heritage	Flora & Fauna	Social & Economic	Landscape	Dust, Noise & Vibration	Road Traffic & Transport	Water Quality
CONSTRUCTION								
Site Access and Infrastructure			•		•	•	•	•
Site Clearance		•	●		•	•	•	•
Raw Materials					•			
Transport of Raw Materials						•	•	
Transport of Employees							•	
Works to River Channel			●			•	•	●
Dams and Diversion Works			•					●
Embankment Construction/Improvement	•	•	●		•	•	•	
Wall Construction / Improvement								
Culvert Construction / Improvement								
Construction of Other Structures or Mitigation Measures								
Waste Disposal								
Site Restoration & Landscaping			o		o	•	•	
OPERATION								
Embankments	o		●	○	o			
Walls								
Culverts								
Other Structures or Flood Mitigation Measures			•		•			

KEY: POTENTIAL IMPACTS

● Major Negative Impact
• Minor Negative Impact
○ Major Beneficial Impact
o Minor Beneficial Impact
No Discernible Impact

81

assessments need to be adapted to local conditions due to the importance of site factors in controlling the pattern, type and frequency of slope failures. However, the general principles of the approach to investigation developed in the study of the landsliding at Ventnor, Isle of Wight have a broader relevance to many coastal cliff environments. There is great potential for adapting this type of study to flood prone areas, to demonstrate how flood risk information can be incorporated into the planning process.

Scoping studies to provide a qualitative assessment of the possible environmental effects of a development project should be based on consultation with the key interest groups and authorities. Environmental receptors can be defined and a matrix established to set out the likely level of impacts associated with different project activities. This type of study can provide a focus for any subsequent environmental assessment, if required.

Chapter 6: References

British Standards Institution 1981. BS5930. A code of practice for site investigations.

Clark A.R, Palmer J.S, Firth T.P. and McIntyre G. 1991. The management and stabilisation of weak sandstone cliffs. In J.C. Cripps and C.F. Moon (eds) The engineering geology of weak rock, 392–410.

Cooke R U and Doornkamp J C, 1990. Geomorphology in environmental management. Oxford University Press.

H.R Wallingford, various dates. Macro Review of the Coastline of England and Wales. 10 Volumes.

Institution of Civil Engineers 1933. Committee on floods in relation to reservoir practice. Interim Report.

Institution of Civil Engineers 1981. Flood studies report – 5 years on. Proc. Conference, Manchester. Thomas Telford.

Lee E.M. and Moore R. 1989. Landsliding in and around Luccombe Village, Isle of Wight. HMSO.

Lee E.M. and Moore R. 1991. Coastal landslip potential assessment. Ventnor, Isle of Wight. Report to DoE.

NERC 1975. Flood Studies Report. Institute of Hydrology.

MAFF 1993. Coastal defence and the environment. MAFF Publications.

Pratt I. 1993. Operation of the Storm Tide Warning Service. Proc. MAFF Conference of River and Coastal Engineers.

Pugh D.T. 1987. Tides, surges and mean sea level. John Wiley and Sons.

7 Detailed Assessment

Introduction

If the desk study or ground inspection indicates that the selected site is likely to be affected by erosion, deposition and flooding processes, then **detailed assessments** should be undertaken by the developer to ensure that satisfactory precautions can be designed to safely overcome the potential problems. In many cases the local planning authority may specify that planning applications for development in vulnerable areas be accompanied by a specialist report describing and analysing the relevant issues and indicating how they would be overcome.

Detailed assessments or investigations should be carried by developers out to enable decision makers to:

- demonstrate to the local planning authority that the proposed scheme design takes full account of the relevant erosion, deposition and flooding issues;

- determine individual planning applications. Permission may be granted, refused or granted subject to conditions specifying the measures to be carried out to overcome potential problems;

- assess the level of risk associated with the development and determine the economic viability of the protect, taking account of the costs of any necessary defence works;

- establish the design criteria for defence works;

- assess the likely environmental effects of the proposed development and associated defence works, through consideration of sediment budget and sensitivity issues.

Sources of Information

Successful design of developments and defence works requires a thorough investigation of the ground conditions and physical processes operating at the selected site **and** the surrounding area. This would typically include a combination of the desk study and site reconnaissance techniques described in Chapters 5 and 6, field measurement, sub-surface investigation, laboratory testing, conceptual modelling, mathematical modelling and physical modelling (Table 7.1). A wide range of investigation techniques can be employed in different settings (see, for example, Table 7.2); the scale of the investigation will depend on the character of the proposed development and the severity of the problems facing the developer and the amount of investigation already undertaken as part of the general assessment or site reconnaissance.

In general terms the investigation will need to establish:

- the foundation conditions at the site;

- the degree of risk from erosion, deposition and flooding;

- the nature and scale of defence works required to reduce the risks to an acceptable level;

- the potential range of schemes that could be used to overcome the problems, identifying suitable sources of materials and their costs;

- the likely effects of the proposed works on sediment budgets, the environment and other interests;

APPROACH	ITEMS
Field Measurement	• erosion and accretion rates • ground movement • topography and slope steepness • river flows • river corridor surveys • sediment transport • coastal zone surveys • beach, mudflat and saltmarsh profiles • waves, tides and currents
Sub-surface Investigation	• foundation conditions • landslide form and geological conditions • groundwater conditions
Laboratory Testing	• particle size analysis • shear strength • consolidation
Conceptual Models	• river channel instability • sediment budgets • landform change • landslide development
Mathematical Models	• hillslope and wind erosion • river flows • flood routing • wave, tides and surge conditions • joint probability assessments • sediment transport • slope stability
Physical Models	• the performance of defence works • the effects of schemes on the environment

• the mitigation measures that might be used to counterbalance any undesirable affects of the works;

• the cost of works relative to the benefits they would produce.

This Report is not the place for a detailed discussion of detailed investigation procedures, as these are readily available elsewhere (e.g. BSI, 1981; Weltman and Head, 1983; Site Investigation Steering Group, 1993). However, five brief case studies will be described to illustrate how various techniques are used in combination to investigate problems in the river environment (the River Sheaf and Bridport Flood Alleviation Schemes), flooding in estuaries (the Thames Barrier), coastal erosion (Monk's Bay, Isle of Wight) and coastal instability (Holbeck Hall, Scarborough).

The River Sheaf Flood Alleviation Scheme

Flooding of the culverted River Sheaf, in South Yorkshire, can be expected one year in five (Young and Cross, 1992). Major events have occurred in 1922, 1958, 1973 and 1982 causing considerable traffic disruption in Sheffield city centre and the flooding of many commercial and residential properties in the city. A flood alleviation scheme was promoted by the Yorkshire Region of the NRA and is intended to improve the defences along a 5.2km stretch of the river to a 1 in 50 year standard, at an overall cost of £3.1M.

Flood flows were determined from the Flood Studies Report (NERC, 1975), although historical data from gauging stations is limited. The regional growth curves (Chapter 6; Figure 6.2) were used to predict high magnitude levels at the nearby gauging station; flood frequency at other points in the catchment were determined using the flood studies ungauged catchment method.

Table 7.2 Typical study components and methods for coastal defence works (after Fleming, 1992).

SUBJECT	METHOD
Deep Water Wave Climate	Field Measurement Numerical Hindcast from Wind Data
Tide Levels and Currents	Field Measurement 2D Numerical Model
Storm Surge	Long Term Water Level Measurements 2D Numerical Model
Shallow Water Wave Climate	Field Measurement Numerical Models – Wave Refraction/Diffraction
Design Criteria	Statistical Analysis of Extremes and Joint Probability Indicative Standard of Protection
Sediment Circulation	Historic Analysis Numerical Model – Tidal Residuals Field Measurements – Sediment Path Analysis
Beach Response	2D/3D Physical Model Numerical Beach Plan Shape Numerical Profile Model Numerical 3D Response Model
Run–Up and Overtopping	2D Physical Model Semi–Empirical Calculations
Structural Integrity	2D/3D Physical Model

Flood levels were calculated using a computer-based **mathematical model**, derived from empirical formulae. Existing weirs were used as control points; losses at bridges, culverts and constrictions were calculated manually. However, the model could not produce reliable results for critical stream lengths and it proved necessary to produce a **physical model** (i.e a scale model of the area of interest made out of suitable materials) of four river sections at a cost of £60,000. The physical models could not be calibrated from historic events because of the lack of reliable data, but observed flow patterns were reproduced accurately in the model. The performance and effect of the various improvement options were tested in the models.

A **river corridor survey** was carried out to determine the environmental issues that needed to be taken into account. These included: river bed and bank habitats, and buildings of industrial archaeological value. Following consultation with the local authority and the conservation bodies, an **environmental assessment** of both the permanent and temporary works was prepared.

As part of the design studies, a detailed underground survey of the **river channel**

sediments and debris was carried out through the culverted sections. The results of this survey indicated that excavation and removal of consolidated silt and stone in the river bed would be a major task. The structure of the culverts was closely inspected to record their condition as it was intended to use a sprayed concrete coating to improve their hydraulics. A detailed sub–surface geological investigation was carried out to enable the detailed design of the automatic debris screens to prevent blockage of the culverted sections.

Bridport Flood Alleviation Scheme

Bridport, Dorset is situated adjacent to the River Brit and is periodically affected by flooding. In May 1979 major flooding (a 1 in 35 year event) occurred with 450 houses and 200 caravans inundated, and the A35 trunk road severed by floodwaters up to 1.2m deep (West and Mann, 1987). An outlined scheme to alleviate flooding to the 1 in 100 year standard was proposed by the Wessex Water Authority, at an estimated cost of £3M. Owing to limited financial resources it was decided to undertake the scheme in six stages over a ten year period.

River flow and rainfall records were insufficient to give more than a general indication of flood conditions. As a result design discharges were calculated using the methods for ungauged catchments presented in the Flood Studies Report (NERC, 1975).

The scheme design was prepared using a mathematical model for flood routing, (FLUCOMP1). This model is a version of **HR Wallingford's** suite of flood routing models that can be used to mathematically model river flows. The modelling demonstrated that although the preferred scheme would pass the design flow through Bridport, there was a much smaller safety margin at West Bridge. In order to confirm the schemes design parameters and to ensure that the most effective works were undertaken, a **physical model** was commissioned from HR Wallingford to cover the reach from Palmers Brewery to North Mills, at a cost of £64,000.

The model of the channel was made at a scale of 1 in 50 from sand cement mortar on well compacted sand and gravel fill. The resulting model was 32m long and 10m wide. About 65 measured cross sections were used to produce an accurate channel section. Bridges and other structures were made from wood and then carefully positioned. The discharge of water from the model was at a scale of 1 to 17,678.

The model was used to simulate flood events; the effects of the addition of trees and hedgerows on the floodplain and riverbank were also modelled. The latter test indicated that if planting was carried out to enhance the nature conservation interest, the flood defences would have to be raised in height, by at least 40mm.

The Thames Barrier

Tidal flooding has long been a problem in the Thames estuary, posing a considerable risk to low lying parts of London. Although flood embankments have been in place for many years, by 1969 the probability of tides equalling or exceeding the flood defence levels as set in 1930 were thought only to be once in ten years. By 1980 this probability was considered likely to increase to once in seven and a half years (Greater London Council, 1970).

In the late 1960's the Greater London Council were charged by the Government with the task of investigating the problems and producing a solution. Relevant studies were commissioned from a variety of research institutes, consultants, university departments and public bodies. HR Wallingford, for example, were requested to carry out studies to predict how the construction and operation of a barrier at different sites would affect river levels, tidal currents and siltation in the Thames. This involved (Kendrick, 1988):

(i) **field measurements**; four **large scale surveys** of the **tidal reaches** were made on spring and neap tides following high and low river flows. Every ten minutes, measurements were made of tidal level, current velocity, salinity, temperature and suspended solids concentration. Smaller surveys were carried out adjacent to the navigation channel to provide more detail in the reaches most likely to be affected by the barrier.

Synchronous **water level measurements** were made at ten minute intervals throughout a tide at sixteen stations to check tidal behaviour against the historical records.

A **sub-bottom survey** was made of the reaches adjacent to the barrier site to determine the thickness of the sediments; this was carried out to assess the potential for river bed erosion and increased sedimentation problems in, for example, dock entrances. Tests were also undertaken to establish silt transport and settlement rates during the tidal cycle. A programme of **continuous silt monitoring** was established to supplement the sediment information; measurements were made by optical sensors fixed at two heights above the river bed on jetties or other structures.

(ii) **laboratory testing**; tests on the behaviour and properties of Thames silt were carried out in flumes and settling columns.

(iii) **computer modelling**; a suspended silt movement model was developed, reproducing the estuary form with the channel simulated by a variable width rectangular section (Owen and Odd, 1970). The model was used to establish how the closing and re-opening of a barrier on successive tides would be likely to affect the distribution of silt in the estuary.

86

(iv)   **physical modelling**; six scale models were tested to examine problems associated with the barrier, the most important being the Thames estuary model, the Silverton barrier model and the Bed Protection model (Table 7.3).

The **Thames Estuary Model** was 115m long and reproduced the tidal reaches from the tidal limit to the sea. It provided information on the behaviour of the river before, during and after construction. The model was tested to examine the effects on water levels, salinities, currents and depth of navigation channels for various barrier options and locations.

The **Silvertown model** reproduced the Thames from the tidal limit to a position 70km downstream and was used to carry out detailed sedimentation studies of the barrier site and adjacent reaches. Wood grains were used as bed sediments and tests were carried out to establish how the scheme affected navigation and siltation.

The **Bed Protection model** was used to investigate the design of the river bed protection works needed to prevent undermining of the barrier.

A mathematical model of the river hydraulics was developed by the Institute of Oceanographics Sciences to complement the physical models. This model was used to study the effects of water spilling over the river banks.

An extensive desk study was undertaken which used archives from the Institute of Geological Sciences, local literature and previous ground investigations to provisionally identify the nature and extent of geological materials present at the barrier site. These studies showed that north of the river, alluvium overlies Thanet Sands and Chalk, whereas south of the river, alluvium lies directly on the Chalk. Preliminary ground investigations were carried out by Foundation Engineering Limited in Woolwich Reach during 1970, who sunk seven boreholes. The detailed investigation consisted of 73 boreholes (54 over water) along the axis of the barrier between December 1971 and May 1972. Standard geotechnical logging and testing was carried out. In addition, seventeen land boreholes were sunk, north and south of the river as part of the groundwater study, in conjunction with the Institute of Geological Sciences work. The basic technique used was shell and auguring into the overlying soils and weaker Chalk and continued by rotary coring where required (Fookes and Martin, 1978).

Piezometers were installed in six boreholes to assist in understanding the influence of the tide in affecting groundwater levels. Sampling of the river bed materials was also undertaken to assess any restraints on the functioning of the barrier by deposition. Sediment traps were also located for up to twelve weeks, and the subsequent captured material subjected to grading analysis. All the field tests were backed up by laboratory testing on recovered samples to determine strength and deformation properties. The ground investigation which cost £108,634 was considered successful in the production of suitable parameters used for the foundation design of the barrier.

Monk's Bay, Isle of Wight

A large coastal landslide in Monk's Bay in Ventnor, Isle of Wight, during the winter storms of 1989/90 and progressive deterioration of Victorian sea walls led to a number of properties being threatened by erosion of the soft Lower Greensand cliffs (Andrews and Powell, 1993). The local authority (South Wight BC) recognised the need for a major coast protection scheme, as continued erosion could lead to the reactivation of the Undercliff landslide complex, inland. The Monk's Bay area had traditionally been an important tourist beach and the authority sought consultants Posford Duvivier to investigate the possibility of including an amenity beach as part of the scheme, using beach nourishment.

A **physical model** was developed by HR to test the preferred scheme combination of an offshore breakwater, groynes and beach nourishment. The model was built at 1 in 80 scale with the sea bed reproduced over an area of approximately 900 x 600m. Random seas were produced by a 10m long wave paddle driven by an electro–hydraulic system, generating undirectional waves. Graded anthracite was used to simulate the mobile beach materials, although the relatively low specific gravity would accelerate the rate of longshore transport; calibration of the model was, therefore, needed with observed rates of transport.

The performance of the beach containment structures was tested in two storm types:

Table 7.3 Thames Barrier physical models (after Kendrick, 1988).

MODEL	HORIZONTAL SCALE	VERTICAL SCALE	ESTUARY LENGTH REPRODUCED (km)
Thames Estuary	1/1600	1/60	100
Inflatable Barrier	1/65	1/65	0.9
Silvertown Barrier	1/300	1/60	70
Barrier Gate	1/25	1/25	0.5
Bed Protection	1/50	1/50	2.5
Outer Estuary	1/1000	1/1000	150

. a frequently occurring event (five times a year) to model patterns of beach development commensurate with longer periods of wave activity;

. severe storm events (one in fifty years) to model extreme levels of beach response.

Holbeck Hall, Scarborough

The major landslide on the cliffs of Scarborough's South Bay in June 1993 destroyed a four-star hotel (Holbeck Hall) and raised considerable local concern for the safety of neighbouring properties. Investigation work began immediately; Scarborough BC called in consultants, Rendel Geotechnics, to assess the situation and design emergency works (Clark and Guest, 1994). Because of the time pressures several lines of investigation were pursued in parallel, including:

(i)   **field measurement**; detailed topographic maps of the site were prepared by the Ordnance Survey from specially flown vertical aerial photographs. A geomorphological map of the slide was produced using these photographs as a base; this map indicated the nature and extent of the individual landslide units and identified areas of seepage.

The surrounding slopes were monitored to detect any further movements. Initially, this was by routine survey of pins placed on the cliff edge and regular inspection of tension cracks. This was later augmented by an automatic sensor system which can detect movements of a fraction of a millimetre

and immediately warn council staff by a radio-pager system.

(ii)   **sub-surface investigation**; twelve boreholes were sunk within the landslide mass to establish the nature of the materials and determine the position of the basal shear surface. Piezometers were installed to monitor groundwater conditions and samples of the materials were taken for **laboratory testing** of the shear strength.

(iii)   **stability model**; eye witness accounts, the geomorphological mapping and sub-surface investigation results were integrated to produce a model of the way the landslide developed from initial small failures to the dramatic and sudden cliff collapse. This model provided a general indication of the mechanism of failure and was used to examine the options for emergency treatment works, including the stable slope angle for regrading works.

Summary

Detailed assessments or investigations will often be necessary to establish the design criteria for defence works, to evaluate the likely performance of design options, to assess the likely environmental effects of the works, and to examine the costs and benefits to ensure that any scheme is worthwhile and the most economic of the feasible options. In many cases detailed information will need to be supplied by a developer in support of planning applications; the local planning authority will determine whether the application should be granted permission, considering the way the proposals have taken account of any potential problems.

A wide variety of detailed investigation techniques are available; in most cases, a combination of techniques will be most appropriate. Investigations should examine the ground conditions at the site and the physical processes operating at the site and in the surrounding area. This can typically include investigating:

- the flood character of a stretch of river or stream, i.e. the magnitude and frequency of events and the routing of floods;

- patterns of river channel instability, erosion and accretion;

- sediment movement through relevant sections of river channels, estuaries or the coastline;

- movement of sediment onshore and offshore;

- the sediment budget for a stretch of coast, identifying source and sink areas;

- the sensitivity of a river or coastal section to change;

- the stability of coastal cliffs, identifying the sensitivity to change in groundwater levels, material strength or slope geometry;

- the environmental effects of the proposed works.

Flexibility is essential for successful detailed investigations, with the range of techniques used specifically directed towards addressing the potential problems that occur at a particular site. Detailed investigation should not be considered as a series of predetermined and isolated steps. The natural variability of both materials and processes in river and coastal environments precludes such an approach, with each site requiring individual assessment. The examples presented in this Chapter illustrate how the nature of an investigation can vary from, for example, flooding to coastal erosion problems, although the objectives will remain the same – to produce the information necessary for an economic and safe design that takes account of potential environmental effects. The success of any investigation is partly dependent on the ability of the engineer to envisage what problems could be encountered; this, of course, means that detailed

investigation should be preceded by a through desk study and other site reconnaissance techniques (see Chapter 6).

Chapter 7: References

Andrews J. and Powell K. 1993. Monk's Bay scheme Isle of Wight. Proc. MAFF Conference of River and Coastal Engineers.

British Standards Institution 1981. BS5930. Code of Practice for Site Investigation. BSI.

Clark A.R. and Guest S. 1994. Holbeck Hall landslide: coast protection and cliff stabilisation. Proc. MAFF Conference of River and Coastal Engineers.

Fleming C.A. 1992. The development of coastal engineering. In M.G. Barratt (ed) Coastal zone planning and management 5–20. Thomas Telford.

Fookes P.G. and Martin P.L. 1978. Site investigation and geotechnical considerations. In Thames Barrier Design, ICE, 29–50.

Kendrick M. 1988. The Thames Barrier. Landscape and Urban Planning 16, 57–68.

NERC 1975 Flood Studies Report. Institute of Hydrology.

Owen M.W. and Odd N.V.M. 1970. A mathematical model of the effect of a tidal barrier on siltation in an estuary. Int. Conf. Utilisation of Tidal Power. Halifax, Nova Scotia.

Site Investigation Steering Group 1993. Site Investigation in Construction. Thomas Telford.

Weltman A.J. and Head J.M. 1983. Site investigation manual. CIRIA Special Publication 25.

West G.M. and Mann K. 1987. Bridford Flood Alleviation Scheme. Journal of the Institution of Water and Environmental Management, 291–296.

Young D.R. and Cross J.G. 1992. Design and Implementation of the River Sheaf Comprehensive flood–alleviation scheme. Journal of the Institution of Water and Environmental Management. 6, 20–27.

8 Acceptance of Risk

Introduction

Voluntary acceptance of some level of risk is inherent in every decision to purchase property, carry out economic activity or provide services in vulnerable areas. The resulting costs that can arise when damaging events occur are borne by the affected parties through **maintenance and repair, emergency action** or offset by **insurance** claims or **legal action**. Frequently the acceptance of risk makes sound economic sense. Indeed, in many instances the benefits of placing property, infrastructure and services in vulnerable areas may outweigh the risks from erosion, deposition and flooding. As was described in Chapter 2, property in vulnerable areas is often cheaper than equivalent property in other areas; here, the savings in property cost can be balanced against the probability of losses or repair costs. Elsewhere, the costs of providing protection measures against, for example, hillslope erosion may considerably outweigh the costs of clear up operations when damaging events occur.

Maintenance and Repair

Individuals are responsible for maintaining and repairing their own property. Much can be done to reduce the effects of flooding on the building fabric and contents; a range of permanent, contingency or emergency floodproof measures are available (Table 8.1), but experience in many flood prone areas suggests that they are not widely practised (e.g. Harding and Parker, 1974). In areas of coastal instability, the effects of ground movement can be reduced by precautionary works and repairs; these measures can reduce maintenance costs and may prolong the life of the property. Neglect can lead to instability and flooding problems at a property and in the surrounding area; maintenance of individual properties and its watercourses can be of great importance, as was recognised in Ventnor, Isle of Wight where a series of suggestions for good practice were prepared for homeowners in unstable areas, (Figure 8.1; Table 8.2).

In areas prone to soil erosion by water or wind, the **highway authority** has a duty to remove soil that obstructs the highway and may recover expenses reasonably incurred from the person responsible for the obstruction (the **Highways Act 1980** S.150). Under S.151 of the Act, the authority may serve a notice on a landowner requiring him to undertake, within 28 days, such work as will prevent soil being washed onto the road. In 1990, for example, the Isle of Wight County Council served notice to 35 farmers/landowners under S.151 of the Act, when intense rainfall caused extensive erosion of agricultural land and flooding; the highway clearance costs were estimated to be about £25,000. The Notice acted as a warning that action would be taken unless the farmers took measures to prevent soil washing onto highways; most farmers cooperated in trying to prevent a recurrence of flooding by the use of straw bales and the digging of ditches.

Highway authorities are also responsible to keep roads, with the exception of trunk roads and motorways, free from flooding (the 1980 Act S.100). It is also the duty of a highway authority to set out posts or stones on roads subject to frequent flooding, indicating the depth of floodwater covering the highway (S.103). The maintenance costs associated with flood problems can be sizeable; Northamptonshire County Council Highways, for example, incurred costs in the region of £27,000 in 1992. Often the maintenance works will involve clearing out drainage gullies, culverts and ditches or repairs to structures.

Drainage authorities (the NRA, local authorities and IDB's in England and Wales) have the powers to carry out maintenance or repair works along designated rivers and streams, (see the

Table 8.1 Structural adjustments as floodproofing measures (after Sheaffer, 1960).

MEASURE	MATERIAL PROTECTED	CLASS OF MEASURE	STRUCTURAL PREREQUISITES
Seepage Control	St–Co	P–C	Well constructed
Sewer Adjustment	St–Co	P–C	None
Permanent Closure	St–Co	P	Impervious walls
Openings Protected	St–Co	C–E	Impervious walls
Interiors Protected	St	P–C	None
Protective Coverings	St–Co	P–C–E	None
Appliance Protection	Co	E	None
Utilities Service	Co	P–C–E	None
Roadbed Protection	St	P–E	Sound structure
Elevation	St–Co	P–C–E	Sound structure
Temporary Removal	Co	E	None
Rescheduling	Co	E	Alternatives
Proper Salvage	Co		None
Watertight Caps	Co	P–C	None
Proper Anchorage	St–Co	P–C	Sound structure
Underpinning	St	P	Sound structure
Timber Treatment	St	P	None
Deliberate Flooding	St–Co	E	None
Structural Design	St–Co	P	Design

St = Structure
Co = Contents
P = Permanent
C = Contingent
E = Emergency

Table 8.2 Precautionary works and repairs for buildings in unstable areas (after Lee and Moore, 1991).

Foundations	Build on rafts to 'float' over slight movement; subdivide rafts into simple rectangular shapes.
Structural Form	Framed construction has better resistance to damage than masonry construction. Allow for future repairs and movement. Allow extended bearings for supports. Consider design features which provide integral buttressing.
Property Walls	Provide movement joints as frequently as possible/practicable. Subdivide complex structures with movement joints.
Freestanding Walls	Provide weep holes at upper and lower levels; build in suitable designed movement joints.
Joinery	Design to provide flexibility with large rebates, dry jointed frames and loose pin hinges. Allow for future use of folding wedges. Generally use accessible screw fixings. Consider glazing with beading, soft mastic or gaskets.
Ceilings/Linings	Consider matchboarding or sheet materials. Cut gap around old ceilings with cove over. Overboard old ceilings. Avoid tight fitting of all cladding; loose fit with removable fixings.
Renderings	Incorporate expanded metal reinforcement.
Gutters	Provide gutters substantially in excess of BRE recommendations to ensure that all rain wall is collected in heavy storms (suggested design standard 125mm/hr).
Hardstanding	Provide frequent waterproof movement joints, perhaps at 2m centres. Subdivide all concrete into rectangular shapes, incorporate light steel mesh reinforcements.
Drainage	Inspect and repair existing drains. Ensure that all areas are properly drained. Landscape surfaces to provide falls towards drain inlets.
Sealants	Almost all sealant materials require periodic renewal; allow for access; choose appropriate materials.

Figure 8.1 Suggested good maintenance practices for homeowners on unstable land (after Lee and Moore, 1991).

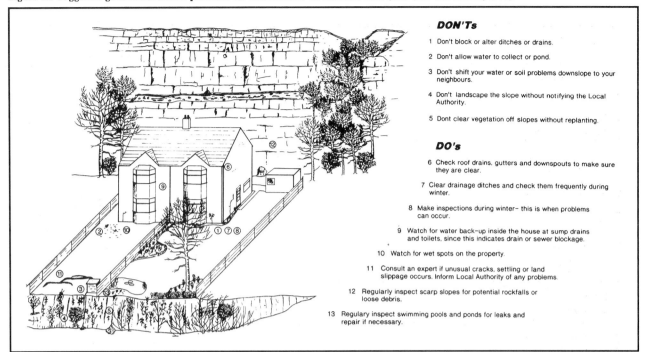

DON'Ts

1 Don't block or alter ditches or drains.

2 Don't allow water to collect or pond.

3 Don't shift your water or soil problems downslope to your neighbours.

4 Don't landscape the slope without notifying the Local Authority.

5 Dont clear vegetation off slopes without replanting.

DO's

6 Check roof drains, gutters and downspouts to make sure they are clear.

7 Clear drainage ditches and check them frequently during winter.

8 Make inspections during winter– this is when problems can occur.

9 Watch for water back-up inside the house at sump drains and toilets, since this indicates drain or sewer blockage.

10 Watch for wet spots on the property.

11 Consult an expert if unusual cracks, settling or land slippage occurs. Inform Local Authority of any problems.

12 Regularly inspect scarp slopes for potential rockfalls or loose debris.

13 Regulary inspect swimming pools and ponds for leaks and repair if necessary.

Methodology Report where responsibilities in Scotland are also described, Rendel Geotechnics). Maintenance is frequently undertaken on a routine basis to reduce the likelihood of flooding. In general, maintenance involves cutting and removal of obstructions, clearance of rubbish from urban channels and the dredging of accumulated shoals. An indication of the costs involved is provided in the NRA's 1992/93 Corporate Plan which lists the average cost of over £1000 per kilometre of main river flood defences maintained (NRA, 1992). Drainage authorities may also serve notice to a landowner to carry out maintenance works if a non main river watercourse is in a condition where the flow is impeded (Land Drainage Act 1991 S.25; Water Resources Act 1991 S.107). Alternatively, the drainage authority may carry out the work itself and recover the expenses from the landowner.

The annual growth of plants is a particularly serious problem in many lowland areas, where abundant vegetation can impede drainage. On the River Frome, in Dorset, it is estimated that the amount of weed removed during an annual cut is between 1.3–2.6 tones per hectare. Plants can be managed by a variety of methods including manual cutting, the use of mechanical weed cutting boats, use of herbicides or by grazing with fish such as carp (e.g. Brookes, 1988). Efficient plant control is also of benefit to anglers and fisheries managers.

Harbour authorities and **navigation authorities** (e.g. the British Waterways Board) need to carry out maintenance works to ensure that navigable

channels remain open for vessels. Dredging is the most important activity and, here, it is important to distinguish between **capital dredging**, for example, to create a new channel (i.e. it costs a single capital sum) and **maintenance dredging** to tackle a recurrent sedimentation problem. Not all rivers and estuaries are seriously affected by sedimentation; the need for dredging will be determined, in part, by the location of the development and facilities (Table 8.3).

A wide variety of dredging methods are available. Extraction of the material involves either **mechanical methods**, such as digging with a bucket or cutting by means of a series of blades, or **hydraulic methods** where water and spoil is pumped from the bed. Dredged material can then be deposited on land or elsewhere in the river or sea:

(i)   **on land**; suitable sites need to be identified with sufficient capacity to contain the material. Important considerations include the effect of the material on slope stability and the potential for environmental pollution problems;

(ii) **in water**; this may involve deposition in either of two types of sediment transport systems:

–   uni–directional systems, such as rivers where dumping downstream will ensure that the material does

Table 8.3 The need for dredging in rivers and estuaries (after Bray, 1979).

PORT AND HARBOUR ACTIVITY	LOCATION
Development without Dredging	• sites on rocky coasts where inshore waters are deep or sheltered bays • sites where development does not impede sediment movement e.g. by using open pile jetties • sites where development helps move sediment away from the site e.g. self cleansing harbours
Development with Capital Dredging only	• sites on rocky coasts or in sheltered bays with insufficient water depth • site which are stable after development; this can occur in rivers with rocky beds and where the enlargement of the cross section does not significantly reduce current velocities
Development with Maintenance Dredging	• sites where there are natural sedimentation problems • sites where the new channel configuration promotes further sedimentation

not return;

– cyclic systems, such as tidal estuaries where it has to be assumed that material will return.

Coastal protection and **drainage authorities** are responsible for maintenance works on coastal defences. Generally such works will involve repairs to storm damage, beach nourishment or reinstatement, cleaning out slope drainage or control of vegetation. In England and Wales they have been advised by MAFF/Welsh Office that all maintenance should be subject to a **maintenance plan** or **post project evaluation exercise** to allow review of maintenance requirements to take place, (MAFF, 1993a).

In general terms, maintenance and repair works can reduce the risk of failure of structures (buildings, bridges or flood and coastal defence works), increase the planned life of the structure or the period of effectiveness of a channel. The range of maintenance works described above can be considered **essential** for many interests, especially ports and harbours or inland waterways. However, maintenance works can lead to conflict with the need to protect the natural environment, especially within river corridors and estuaries. When disputes arise the mechanisms for resolving conflict involve consenting procedures or dialogue with statutory or recommended consultees, as illustrated by the following examples.

(i)   **Dredging**; although, many dredging operations are authorised by statutory

provisions or have general planning permission in advance (permitted development) they are effectively constrained by the need to find safe sites for the disposal of dredged sediment. Problems are enhanced when such sediment is contaminated by heavy metals or other potentially harmful substances that, in the past, had been freely discharged into the waters. British Waterways, for example, has been undertaking an environmental dredging programme (Project Aquarius) to remove contaminated silt from canals.

Consent for disposal of dredged material is required from either the relevant waste regulation authority or MAFF/SOAFD for land or sea based disposal, respectively. Discharge consents may also be required from the NRA when the proposed dump site is in inland or controlled waters. As environmental considerations are an important factor in determining whether to issue the appropriate consents, navigation and harbour authorities may find themselves in the position of needing to dredge to remain competitive and finding disposal options limited by potential pollution problems that are a legacy of past industrial practice.

In considering an application for a licence to dispose of dredged material at sea, the Licensing Authority (MAFF or SOAFD) is required (under The Food and Environment

Protection Act 1985 Part II) to have regard to the need to:

- protect the marine environment;
- prevent interference with other legitimate users to the sea;
- the practical availability of any alternative methods of disposal.

The consideration of alternatives to disposal at a dump site can be problematic. Ash (1994) has identified four broad categories for the **beneficial use** of dredgings, including coastal defence (e.g. beach nourishment), habitat creation, land reclamation for development and retention within the coastal system to maintain the sediment budget. Table 8.4 highlights the beneficial uses of dredgings from the expansion of the Port of Felixstowe in the early 1990's, with 760,000m^3 of coarse material used in NRA coastal defence schemes and 400,000m^3 used for habitat creation. However, the majority of dredged material is fine, the properties of which are poorly understood and for which beneficial uses are harder to find, especially if the sediment is contaminated by heavy metal pollutants.

The beneficial use of dredged material can help ensure that deposition within estuaries is not a completely negative process. However, effective use of the material depends on coordination between port and harbour authorities, coast protectional authorities, the NRA and conservation bodies. Here, the recently established **coastal defence groups** have a key role (see Figure 3.2).

(ii) **Maintenance of flood and coastal defences** can also have implications for conservation interests. Weed cutting and dredging are likely to have significant impacts on fish populations by disrupting feeding, reproduction and normal behaviour patterns. Plant habitats may also be affected by maintenance works (Brookes, 1988). Conservation guidelines have been prepared, however, which advise drainage authorities of the need to consult with conservation agencies and other interest groups before undertaking their annual maintenance programme (MAFF/DoE/WO, 1991). These guidelines emphasise:

- the need for agreement on the frequency, timing and practice of operations;
- heavy maintenance works may need special consideration akin to capital works;
- routine maintenance should be planned to take place outside the normal breeding season for water based or river bank wildlife;
- flowering and seed setting periods of wild plants should be taken into consideration.

Conflict may also arise when maintenance works may lead to the degradation of sites of geological or geomorphological interest, as has occurred along the stretch of the River Dee in Wales, between Holt and Worthenbury (see Rendel Geotechnics 1995). The Dee is one of only a few large British upland rivers with a well developed, meander belt in its lower course that is relatively free from direct human intervention. The reach includes a range of meander forms with numerous cut-offs and abandoned channels. The Countryside Council for Wales (CCW) has proposed to recommend to the Secretary of State that the stretch be designated as an SSSI. River channel migration and regular flooding have, however, been seen as a problem by the local farming community. As part of their flood defence responsibilities the NRA and its predecessors have constructed bank stabilisation works including toe reinforcement using boulders, stone pitching across the bank faces and, more recently regrading and strengthening using geotextiles. However, concerns have been expressed that the measures are likely to conflict with the conservation objectives of CCW.

As part of the SSSI notification procedure CCW must specify those operations likely to damage the special interest of the river corridor. Once notified, owners and occupiers must give CCW written notice before carrying out a potentially damaging operation. The operation can be carried out if CCW gives written consent or if it is carried out in accordance with a management agreement with CCW or if

Table 8.4 Beneficial uses of dredged material from the port of Felixstowe (after Ash, 1994).

1.	Sea Defence bund at West Parkeston for the National Rivers Authority, as part of a grant–aided capital scheme. (250,000m^3 gravel).
2.	Development land for Sea Containers Limited (200,000m^3 of sands and gravels).
3.	Low water gravel berms for foreshore stabilisation at Trimley for the NRA (50,000m^3 gravel).
4.	Low water berm of rock and clay plus beach replenishment at Fulton Hall Point for the NRA (up to 100,000m^3 total).
5.	Foreshore replenishment at Horsey Island for the NRA (100,000m^3 sands and gravels).
6.	Low water berm to provide trickle recharge at Naze North for the NRA (50,000m^3 sands and gravels).
7.	Construction of rock groynes for The Naze coast protection scheme for Tendring District Council (10,000 tonnes).
8.	Creation of a crustacea habitat in a natural deep trench using rock and clay (400,000m^3).

three months have elapsed from the giving of the notice of the operations (**Countryside Act 1968** S.15). If the CCW considers the operation unsuitable, they will either persuade the owner or occupier not to proceed or negotiate a management agreement. Should this approach fail, it may be appropriate for CCW to apply to the Secretary of State for a **Nature Conservation Order** under the 1981 Act S.29. In exceptional cases the site may be compulsorily purchased. There are two exceptions to the requirements for consultation: emergencies and where the operation is authorised by planning permission. The NRA maintenance works on the Dee Meanders are **permitted development** and, hence, notification would not necessarily apply. However, an attempt to resolve the conflict has involved consultation between the NRA, CCW and landowners, seeking to agree the most appropriate management strategy for the river.

The role of the planning system in reconciling the potential conflicts between maintenance dredging or flood defence maintenance works is limited as both classes of operation are permitted development under the General Development Order (GDO) 1988 and, as such, do not require planning permission from the local planning authority:

●   Part 15 Class A (c); Development by the NRA for the maintenance or improvement of any watercourse or land drainage works;

●   Part 17 Class D; the use of **any land** by navigation authorities for the spreading of any dredged material.

A planning authority may, however, make a **direction** under Article 4 of the GDO to withdraw these rights, requiring the operator to seek planning permission. If planning permission is refused, those with an interest in the land could claim compensation under the Town and Country Planning Act 1990 (S.107 and S.108) for any financial loss or depreciation in the value of the land. The option is not often used by local planning authorities for river or coastal management purposes.

Local planning authorities can, however, help ensure that drainage operators or navigation authorities can get access to watercourses to carry out necessary maintenance works. Hereford and Worcester County Council have, for example, included a policy in their Structure Plan which is aimed at ensuring that development makes allowance for adequate access to the river channel for future maintenance or improvements. In St Albans, conditions are attached to planning permissions to ensure that strips are provided along the main river watercourse and kept free from development in order to allow for dredging and maintenance. The importance of such planning policies has been emphasised by the NRA in their guidance notes for local planning authorities (NRA, 1994).

Emergency Action

Peace time emergency planning is not a statutory duty for local authorities, although they invariably take a lead role in many flood or coastal erosion disasters (Table 8.5). Under the Local Government Act 1972 a local authority has the permissive powers to:

- incur expenditure which in their opinion is in the interests of their area or its inhabitants (S.137);

- incur such expenditure as they consider necessary in an emergency or disaster involving destruction of or danger to life or property or where there are reasonable grounds for preventing such an event (S.138(1));

- make grants or loans to other people, bodies in an emergency or disaster (S.138(1)).

The emergency services can be involved throughout a flood event from the preparation of emergency plans to the supervision of recovery operations (Parker, 1988). Table 8.6, for example, summarises the emergency responses during the 1982 York and Selby flood.

Emergency relief ensures that some of the immediate losses are spread throughout the community. The costs of the operation can be considerable; the response to the 1982 York and Selby floods cost a reported £366,000 (Parker, 1988). Rehabilitation costs can be eased by disaster funds or grant payments made **to the local authority** under a "Bellwin" scheme, established under the Local Government and Housing Act 1989 S.155 by which the Department of the Environment/Welsh Office makes available financial assistance for 85% of local authority expenditure incurred above a threshold level (DoE, 1993):

- in providing relief or carrying out immediate works to safeguard life or property, or prevent suffering or severe inconvenience;

- as a result of the incident specified in the scheme;

- on works completed before a specified deadline (usually within two months of the incident).

In exceptional cases, Ministers may be able to announce the approval of a scheme shortly after an incident. However, most localised events would only be eligible for financial assistance after detailed information on the local authority's expenditure had been forwarded to the Department.

However, as was found in the north Wales floods of 1990, difficulties can be experienced in providing financial assistance to all of the most affected victims. After these floods the Clwyd County Council Emergency Planning Team coordinated with the emergency services in providing immediate relief to the flood victims. The police, supported by air force helicopters, RNLI lifeboats, the army, the fire and ambulance services and volunteer lorries, conducted the evacuation of affected areas. Victims were accommodated in ten emergency centres and, later, temporary accommodation. A fund was set up by the Mayor's of Colwyn, Rhuddlan and Delyn to alleviate the hardship faced by many residents in the affected area. The categories of hardship were:

- people who were uninsured;
- people who could prove they were under-insured;
- small businesses;
- fully insured people who sustained uninsured losses, (e.g. fences, gardens etc.).

Around £1.2M was raised, with £813,000 paid out to individuals and businesses by December 1991. The Welsh Consumer Council (1992) reported that problems arose in the distribution of relief as "hardship" was judged according to whether or not a person was insured; no account was taken of the benefits available from other sources including the Department of Social Services. For example, people who were in receipt of income support were also entitled to community care grants worth up to £3000. Those who did not qualify for income support, such as many living on state pensions, received no assistance above the £1000 provided to all hardship cases from the Mayors' fund (Welsh Consumer Council, 1992).

Insurance

Insurance is designed to mitigate the effects of damaging events, but it may not do very much to

Table 8.5 The principal services employed in flood emergencies in Britain.

Statutory Services	
Local Authorities District Councils	Chief Executive Engineers' Department Environmental Health Police
County Councils	Fire Service Emergency Planning (i.e. civil defence) Social Services Education Department Highways Department
Area Health Authority Armed Services	Ambulance Service Military Aid to Civil Communities (MACC) Scheme
NRA	Drainage Engineers

Voluntary Services
British Red Cross Society Women's Royal Volunteer Service Salvation Army Others (e.g. Lions Club)

reduce the vulnerability to that event. The basic aim of insurance is to spread the risk of loss over as large a population as possible and amongst as varied a range of potential losers as possible (i.e. the losses of the few are paid out of the premiums of the many).

An insurance contract is between two parties, the insurer and the insured, and the policy is the written evidence of the terms of contract. These contracts are subject to a great deal of specialised law established in decided cases. Insurance practice tends to work on the basis of a number of assumptions (which may become conditions in certain instances), these are:

- the insured event has a statistical frequency and magnitude which will allow an analysis to be made of the chance or an event of a certain magnitude happening with a defined frequency;

- the damage sustained must be measurable;

- the risk must be spread across both hazardous and non-hazardous places;

- the amount of damage must be limited.

Although insurance against building damage caused by "landslip" has been available since 1971, most companies include clauses in their policies which **exclude cover** for damage due to coastal or river erosion. As a result, insurance is not a viable option for spreading the costs of river-bank or coastal cliff landslides as erosion is invariably amongst the causes of the instability. The availability of flood insurance is, however, more widespread, based on agreements between the industry and the Government made in 1961 (See Rendel Geotechnics, 1995). In these agreements it was indicated:

> "to H.M. Government that the broad effect of the assurance would be that Flood cover would be available from the insurance market to everybody in respect of the contents of his private dwelling – provided it was permanently occupied – at reasonable rates, ... with a small excess. The view was expressed that it would only be in quite exceptional cases that rates in excess of that figure would be charged and that the number of such cases would be very small indeed." (British Assurance Association, 1961)

On this basis, flood damage insurance is available to domestic property holders, and it is a normal part of both the **buildings** and the **contents** policies. The agreement extends to the contents of premises occupied by small traders, so long as that business is carried on throughout the whole year. It was recognised, however, that the premium structure would have to take account of the degree of risk: "in the more vulnerable areas premiums will be higher according to the degree of risk." (See British Insurance Association (1961) Press Release, reproduced in Rendel Geotechnics, 1995.

In practice there are two strands to the insurance of flood risk. The first is the **primary_insurance company**, with whom the policy holder deals, the second is the **reinsurance_company** that will carry the excess of loss in respect of a catastrophic flood event which leads cumulatively to millions of pounds worth of damage. The normal practice is the direct insurer to carry the first £xmillion of claims (plus, in some cases, a percentage of the risk) and for reinsurers to bear the cost of any excess. In practice this means that the direct insurer has to bear the claims costs of comparatively minor incidents whilst the reinsurer becomes involved in the event of a catastrophic flood.

Table 8.6 Emergency service response and costs during the York and Selby floods (after Parker, 1988).

SERVICE	COMMENT
York City and Selby District Councils	The councils brought into operation their flood emergency plans. Large numbers of staff were employed in sandbagging and in cleaning mud and debris. The environmental health departments deployed four teams of 10–20 people to disinfect and dry out houses, jet drains and sewers, remove damaged property and to test for contaminated food.
North Yorkshire County Council	Social workers supervised evacuation of flood victims and coordinated an emergency meals service.
Police	Immediately before the flood, 25 officers were committed to monitoring river levels and informing residents of danger. During the flood event 140 officers were deployed and twenty officers were committed to traffic control during clean up operations. Forty police vehicles were utilised and about 5000 additional miles were travelled during the emergency.
Fire Services	The service received sixteen calls for assistance from the public.
Ambulance Service	Ambulances were deployed during the five days of the flooding, specifically to assist in the emergency.
Yorkshire Water Authority	The authority deployed fourteen staff and 55 workmen; between 6 and 8 January, this increased to seventeen staff and 90 workmen; and between 9 and 26 January ten staff and forty workmen were committed – a total of 8600 man hours including 3200 hours of overtime. These men were primarily involved in strengthening flood embankments and pumping floodwater from behind overtopped defences.
The Army	A labour force of 200 men were supplied from nearby bases to assist operations in York. Helicopters were used to supply food to farms and thirty men were deployed in surrounding communities to take flood victims by boat. In Selby 400–500 men were deployed for about seven days to help with sandbagging and clearing up operations.
Voluntary Services	In Selby, the Women's Royal Voluntary Service (WRVS) was placed on standby on 5 January and provided blankets and portable beds but their main involvement came when the floodwaters subsided and houses needed cleaning up. This cleaning up continued into April. The WRVS was also involved in administering the flood relief fund including the delivery and collection of application forms.
Salvation Army	Visited elderly people and helped with evacuation. They also supplied food to other emergency service personnel.

The terms of flood policies vary according to the situation of the property; premiums in areas of high flood risk can be substantially higher than in other areas. Because insurers generally work on a postcode basis for setting premiums there are instances where some of the residents in a particular postcode are experiencing higher premiums as a result of those few properties which are deemed to be at risk from flooding. Whilst this goes some way to meet the principle of insurance that the losses of the few should be shared amongst the many, it does, nevertheless, seem "unfair" to those who are on risk free sites and yet are faced with those higher premiums. To some extent the insured may benefit from "shopping around" in that the perceptions of risk may differ from one insurer to another. On the whole, however, the insurer (who ultimately has to be concerned with returning a profit on his portfolio) will levy higher premiums in areas of greatest perceived risk. This can raise the cost of property ownership in flood prone areas, and in some cases may lead to prohibitively expensive insurance. The result may be that the occupier fails to take out **contents insurance**, and thereby stands the loss in the event of a flood. As for **building** insurance, however, the situation is different. If a mortgage company is involved, buildings insurance is compulsory and the effect of high premiums is to place a burden on the occupier during their residence, and to make resale of the property difficult (or even impossible).

Amongst domestic property insurers there is a desire to remain competitive in areas of minimal risk whilst at the same time not acquiring over-exposure in areas of greater risk. This same level of competition is not present in respect of the insurance of properties owned by local authorities. Most such properties were insured with the Municipal Mutual Insurance, but since this company was taken over by Zurich Municipal Insurance in 1994, all of that business now lies with the Zurich. Under the terms of their "SELECT" policy, cover is provided for flood damage excluding:

DAMAGE or CONSEQUENTIAL LOSS:

(a) attributable solely to change in the water table level;

(b) caused by frost, subsidence, ground heave or landslip;

(c) in respect of movable property in the open, fences and gates.

... each claim arising (from flooding) will be subject to an EXCESS, the amount of which is specified in the Schedule. Unless stated otherwise, the EXCESS will apply to each and every loss in respect of each separate PREMISES."

This wording essentially follows guidelines laid down by the Association of British Insurers.

Under-insurance of property and contents can be a major problem in flood risk areas. Of 2,800 properties affected by the 1990 north Wales floods, between 1,200-1,500 were under-insured, for a variety of reasons:

(i) the use of policies that were not indexed linked to inflation;

(ii) the use of indemnity policies which discount the value lost through wear and tear. In the event of loss or damage a policyholder would be offered the value of a second-hand replacement rather than a new one;

(iii)the under-valuing of property, especially the value of contents. Problems can also arise when attempting to assess the value of a house for insurance purposes as it is the cost   of rebuilding that is relevant not the market value. Here the **Association of British Insurers** (1991) booklet "Buildings Insurance for Home Owners" provides guidance on rebuilding costs.

In some cases, the decision not to insure adequately is borne out of poverty. Indeed, lack of insurance in the areas of north Wales affected in 1990 was attributed, in part, to the Government's decision in 1988 to remove allowances for insurance from DSS benefits, including state pensions, (Welsh Consumer Council, 1992). In

others, the decision not to take out flood insurance may be rational, albeit based on a lack of awareness of the degree of risk in an area and the consequences of a major event.

Litigation

Litigation is a further means of alleviating the cost of a damaging event. Common law can be used for dealing with those issues not covered by specific legislation; this involves liability for naturally occurring conditions and the questions of natural rights, negligence (duty of care) and right to protection from nuisance. Common law can be extremely complex and the following sections serve only to illustrate some of the key aspects, rather than providing a definitive statement.

If a landowner's natural rights are contravened, this can represent a nuisance. The law of 'nuisance' also protects land from other activities on adjoining properties such as excessive smell, noise or even a growth of tree roots (ref. *Davy v Harrow Corporation (1958), QB 60*). A landowner, therefore, must ensure that his neighbours natural rights are not contravened. Beyond that, however, according to a 1929 judgement in *Pontardawe RCD v Moore-Gwyn (1929) Ch. 656*, as long as a landowner is using his land in an ordinary manner then he is not liable for damage that occurs on adjoining property.

The judgement in the Pontardawe RDC case, however, has been modified by subsequent cases. In *Goldman v Hargrave (1967) 1 AC 645* the landowner was deemed responsible for the damage caused to an adjoining property by a fire, started naturally by lightning. The Privy Council decided that in relation to both natural and man-made hazards the landowner or occupier was under a general duty to his neighbour to take reasonable steps to remove or reduce such hazards if he knew of the hazard and of the consequences of not reducing or removing it. This was explained in a statement by Lord Wilberforce who said:

"The owner of a small property who has a hazard which threatens a neighbour with substantial interests should not have to do so much as one with larger interests of his own at stake and greater resources to protect them."

Leakey v National Trust for Places of Historic Interest or National Beauty (1978) 2 WLR 774, in conjunction with the subsequent ruling by the

Court of Appeal *(Leakey v The National Trust (1980) 1 QB 485)*, provides a clear statement to date on the landowners responsibility for a natural hazard. The case concerned a slope failure in a mound located on National Trust land called Burrow Mump. Natural erosion of Burrow Mump over a number of years had led to 'soil and rubble' falling from the mound onto land owned by the plaintiffs and threatening their houses. The plaintiffs accordingly brought an action in nuisance calling for an abatement of the nuisance and for damages. In 1978 the court decided in favour of the plaintiffs but the defendants chose to appeal against the decision.

The 1980 appeal by the National Trust was dismissed because the court felt that an occupier of land owed a general duty of care to a neighbouring occupier in relation to a hazard occurring on his land whether such a hazard was natural or man-made. This is a fundamentally important decision as far as landslides and landslide hazards are concerned, not least because it arises from a case of slope failure. The general duty referred to the judgement was held to be:

"... to take such steps as were reasonable in all the circumstances to prevent or minimise the risk of injury or damage to the neighbour or his property of which the occupier knew or ought to have known."

The 'circumstances' in this case being described as including:

"... his knowledge of the hazard, the extent of the risk, the practicability of preventing or minimising the foreseeable injury or damage, the time available for doing so, the probable cost of the work involved and the relative financial and other resources, taken on a broad basis of the parties."

In this case it was decided that the National Trust did have the resources to meet the financial burden of the repair work, although clearly any subsequent cases would have to be assessed on their merits.

The question of the resources of the defendant was also broached by Lord Justice Megaw in the Appeal Court where he emphasised that the cost of the works must be considered when deciding whether or not the owner of the land which is causing the danger had discharged his duty of case. Megaw said:

"Take by way of example, the hypothetical instance of the landowners through whose land a stream flows. In rainy weather it is known the stream may flood and the flood may spread to the land of the neighbours. If the risk is one which can readily be overcome or lessened – for example by reasonable steps on the part of the landowner to keep the stream free from blockage by flotsam or silt carried down, he will be in breach of duty if he does nothing or does too little. But if the only remedy is substantial and expensive works, then it might well be that the landowner would have discharged his duty by saying to his neighbours, who also know of the risk and who have asked him to do something about it,

"You have my permission to come on to my land and to do agreed works at your expense",

or it may be,

"... on the basis of a fair sharing of expenses".

In most legal actions concerning erosion, deposition and flooding problems, there are difficulties in attributing responsibility for the damage. For example, to be able to show that a farmer had been negligent in allowing runoff to leave his land and inundate that of neighbours it is necessary to prove that he, or a reasonable farmer in his place, ought to have foreseen the risk of flooding. In this context, an interesting recent case was that of the flooding of a farmhouse and vineyard at Breaky Bottom, near Lewes in East Sussex during October 1987.

In most instances the difficulty of proving foreknowledge of the risks, plus the cost for individuals of taking action against insurance companies, means that plaintiffs are advised not to proceed. In the Breaky Bottom case several circumstances indicated that the action was likely to be successful:

● the farmhouse had been flooded on at least two previous occasions;

● the defendant could be shown to have taken part in discussions concerning the risks of erosion and flooding on the Downs prior to 1987;

● there was no doubt that runoff and sediment came from fields farmed by the defendant;

- there was in existence a local ADAS Advice Note regarding the risks of erosion and specifically of growing winter cereals on steeply sloping land – as at Breaky Bottom.

There is a general perception in the farming community that erosion and flooding results from exceptional rainfall which can be construed as an 'Act of God'. In the Breaky Bottom case legal advice given to the plaintiff was that rainfall events with return periods of 20–25 years could be argued to be within the bounds of what a reasonable farmer would have to taken into account in any land use or management decisions. Central to the plaintiff's case was the contention that the way in which the land was farmed was more important than the magnitude of the rainfall as an explanation of the flooding that occurred.

The Breaky Bottom case was settled out of court and so, there was no legal judgement on the complex issues. As the causes of natural hazards become more clearly understood and specific management advice is given to landowners, it is possible that further instances may come before the courts.

Summary

Who pays when property is damaged by erosion, deposition and flood events? For an individual the choices are clear; bear the loss directly through repairs, take out insurance in anticipation, seek redress through the courts or hope for financial assistance from disaster relief funds. Each course of action may be fraught with difficulties as the reported experiences of the victims of the 1990 north Wales floods have shown (Welsh Consumer Council, 1992); immediate relief may be on hand from the emergency services, but the long process of rehabilitation and reconstruction can lead to intense frustrations.

All these difficulties must, of course, be balanced against the perceived and real benefits of living in vulnerable areas. Accommodation may be cheaper than in "safer" areas and the location can be scenically attractive. It should be noted that in many cases the risks associated with such locations were readily accepted, assuming that the people were aware of the situation.

For highway authorities, coast protection and drainage authorities the costs of maintenance and repair works are met by specifically allocated resources, indirectly paid for by the local residents through their community charges or from central government funds. For bodies such as navigation or harbour authorities, the costs of dredging and spoil disposal are a surcharge on top of their normal operating costs and, by this measure, are a constraint to their efficiency or competitiveness.

Maintenance works can lead to conflict with conservation interests, as highlighted by the dispute between the NRA and CCW over the management of the Dee meanders in Wales and the need to find suitable sites for disposing contaminated silt dredged from navigable waterways. The resolution of these conflicts generally requires dialogue between the various interested parties to seek alternative beneficial uses or acceptable compromises, although the latter may lead to the operating authorities bearing additional costs.

Chapter 8: References

Ash J. 1994. The beneficial use of dredgings in the marine environment. Proc. of the MAFF Conference of River and Coastal Engineers, July 1994.

Association of British Insurers, 1991. Buildings Insurance for Home Owners. ABI.

Bray J. 1979. Dredging: a handbook for engineers. Edward Arnold Press.

British Assurance Association 1961. Flood Insurance Facilities. Letter to Members of the Association Sectional Committees, 2.8.1961.

Brookes A. 1988. Channelized rivers: perspectives for environmental management. John Wiley and Sons.

Department of the Environment 1993. Emergency financial assistance to local authorities: guidance notes for claims.

Harding D. and Parker D.J. 1974. Flood hazard at Shrewsbury, UK. In G. White (ed) Natural Hazards: Local, National and Global, 43–52. Oxford University Press.

Lee E.M. and Moore R. 1991. Coastal landslip potential assessment Ventnor, Isle of Wight. Report to the Department of the Environment.

MAFF/DoE and Welsh Office, 1991. Conservation guidelines for drainage authorities.

MAFF and Welsh Office 1993. A strategy for flood and coastal defence in England and Wales. MAFF Publications.

NRA 1992. Corporate Plan 1992/93.

NRA 1994. Guidance notes for local planning authorities on the methods of protecting the water

environment through development plans. NRA
Bristol.

Parker D.J. 1988. Emergency service response and
costs in British floods. Disasters, 12, 50–69.

Rendel Geotechnics, 1995. Erosion, Deposition
and Flooding in Great Britain. Methodology
Report. Open File Report held at the DoE.

Sheaffer J.R. 1960. Flood proofing: an element in
a flood damage reduction program. Univ. of
Chicago, Dept. Geography. Research Paper No. 65.

Welsh Consumer Council 1992. In deep water: a
study of consumer problems in Towyn and Kinmel
Bay after the 1990 floods. Welsh Consumer
Council.

9 Avoiding Vulnerable Areas

Introduction

An obvious response to erosion, deposition and flooding problems is to avoid those areas where there is an unacceptable level of risk. Individuals are free to make their own choices about where to purchase property; the principle of **caveat emptor** ("buyer beware") dictates that the purchaser is responsible for making sure that the property is safe. Mortgage lenders and insurance companies are also free to make their own judgements about whether to lend money on property or set premiums that reflect the risks in vulnerable areas. This is a very complex situation as greater awareness of problems in vulnerable areas can lead to a general reluctance to purchase or invest in property, resulting in a loss of confidence in the area. Clearly a balance needs to be made between taking on an unacceptable risk and being overcautious.

A more strategic approach is available for ensuring that the potential losses in these areas are not increased by the siting of new development. Here, local authorities have an important role through the operation of the planning system by restricting the development of vulnerable areas. The planning system may also have a positive role in risk management through the discontinuation of land uses or compulsory acquisition of land in vulnerable areas, and providing compensation to the affected parties. Managed retreat from existing lines of coastal defences has become a shoreline management option that needs to be considered in some rural areas.

Restricting the Development of Vulnerable Areas

The planning system provides an opportunity for ensuring that development takes account of the potential problems associated with erosion,

deposition and flooding. This does not mean, however, that development should be prevented in all vulnerable areas, rather that it is suitable and that satisfactory precautions have been taken. Where the risks of damage are high and the provision of new or improved defence works is uneconomic or environmentally unacceptable, preventing the development of the area may be the appropriate response.

Development plans can be used to set out broad strategic policies (Structure Plans or UDP Part I's) or detailed policies (Local Plans or UDP Part II's) that establish a framework for restricting built development. The **allocation of land** for specific types of development can be made with the need to avoid certain vulnerable areas in mind. Here, it is important that the local planning authority has established in advance whether an area under consideration for allocation can be satisfactorily defended (see Chapter 6.) The **development control** process can ensure that planning permission is refused in vulnerable areas. Tables 9.1 and 9.2 provide a range of examples of policies that have been prepared to restrict development in vulnerable areas, in England and Wales, where guidance from the DoE and other Government departments clearly identifies the need for local planning authorities to take account of flooding and coastal erosion problems (see Chapter 3). Typical approaches that have been used include:

(i)   identification of floodplain or coastal lowland areas at risk from NRA S.105 maps (or earlier S.24 maps provided by the former Water Authorities);

(ii)   defining a "set-back line" within which development could be affected by coastal erosion over a particular time period. This is not a straight forward task as coastal erosion rates are difficult to predict over the expected lifetime of a building (see Chapter 5).

Table 9.1 Strategic planning policies for avoidance of flooding and coastal erosion areas, in England and Wales.

PROBLEM	STRATEGIC POLICY
Flooding	Buckinghamshire CC; development should not normally take place in areas liable to flood if it would increase the number of people or property at risk. Clwyd CC; liability to flooding will be considered during the allocation of land. Cumbria CC; development not normally permitted where it would be at risk from flooding. Kent CC; new residential development will not normally be permitted in areas at risk to tidal flooding. Lancashire CC; major development will not normally be permitted in areas at risk from long term flooding (1 in 200 years). Norfolk CC; presumption against new development or intensification of existing development in washlands and floodplains. West Sussex CC; new development or significant intensification in areas at risk will be resisted.
Coastal Erosion	Isle of Wight CC; development of areas of known land instability will not normally be permitted. Norfolk CC; presumption against new building in areas likely to be affected by erosion within the expected lifetime of the development. Suffolk CC; development not acceptable which would be likely to be affected by marine erosion in its lifetime.

Avoidance of risks may also involve imposing constraints on the occupancy of houses in flood risk areas. Chichester DC, for example, prohibit the occupation of all holiday caravans and chalets between 31 October and 1 March when coastal flood risks are greatest. In different circumstances, Brent BC have developed a policy of refusing the creation of habitable rooms in basements subject to flooding.

Similar results can also be achieved through planning policies designed to prevent the spread of development into the unspoilt countryside, urban river corridors or sites of nature and geological conservation value. In this context, it is important to note that not all vulnerable areas have been developed or experience significant development pressures. On the undeveloped coast, for example,

strong protection against unsuitable development exists through statutory conservation designations or non-statutory definitions and the related planning policies. For example, the South Glamorgan coastal cliffs are prone to rockfalls and landslides with average erosion rates of between 0.3m and 0.7m per year; here, Vale of Glamorgan BC have no specific hazard-related policies, but the cliff tops have been designated as a coastal conservation zone where development would not normally be permitted. Chelmsford BC, for example, have set out a policy to maintain the open character and attractiveness of their river valleys to conserve and enhance their landscape value; development is not permitted which is appropriate or insensitive.

In Scotland, the absence of specific planning guidance does not mean that the planning system

Table 9.2 Detailed policies for avoidance of flooding and coastal erosion areas, in England and Wales.

PROBLEM	DETAILED POLICY
Flooding	Bournemouth BC; development not permitted in the floodplain. Chelmsford BC; development within the floodplains of watercourses will not normally be permitted. Chichester DC; permission will not be given for new development in areas at risk. Delyn BC; a presumption against development which would be likely to be subject to flooding. North Norfolk DC; presumption against new development or the intensification of existing development in areas at risk of tidal flooding.
Coastal Erosion	Canterbury City C; the council will safeguard the defined coastal protection zones from development. North Norfolk DC; presumption against new development or intensification of existing development in areas at risk. Suffolk Coastal DC; presumption against development at identified sites. Swale BC; no new development permitted within a development line.

cannot be used to ensure that development avoids particularly vulnerable areas. Whilst the lack of guidance may have contributed to a lack of development plan policies in respect of flooding in Scotland, recent events have instigated a re-appraisal of how local planning authorities address flooding issues. The 1990 and 1993 Tayside floods have heightened awareness of the need, in some areas, to consider flooding as a material planning consideration. For example, Highland Regional Council have recently published a mission statement on flooding and development:

"Recent events around Perth have amply demonstrated the devastation, disruption and expense associated with large scale flood incidents. Many parts of Highland Region have also experienced very high storm and melt water levels during January – the third such occurrence in four years. The usual statistical probabilities of 1 in 50 or 1 in 100 year flood events have little meaning when this kind of pattern occurs. It certainly adds to widespread unease about "global warming", and raises a number of planning issues.

"The risk of inundation to a new development occupying sites on or close by known floodplains is a material planning consideration. In addition to the obvious human risks, the Regional Council also takes on a statutory responsibility to 'defend' communities from flooding. Raising flood banks, culverting watercourses, armouring faces and building holding ponds all consume scarce public resources which might be better deployed elsewhere.

"Since 1990, the Council has been defining areas known to be at risk of flooding in their Local Plans for different parts of the Region. The accompanying policy indicates a presumption against development in these locations, except where this is essential for continued agricultural occupancy of the land. A small step towards the planning of whole river basins may be, but a sensible precaution nevertheless. " (Highland Regional Council 1993).

Following the 1993 floods both Tayside Regional Council and Perth and Kinross District Council produced policies for addressing the flood problems:

(i)   **Tayside RC**; a policy drafted in June 1993 set out a presumption against development in areas liable to flood;

(ii)   **Perth and Kinross DC;** an approved policy on housing in the countryside (March 1993) states that applications for houses will be refused if they are within a flood risk area as defined by the District Council (based on surveys of flood limits).

A number of factors are likely, however, to constrain local planning authorities in Scotland should they wish to be more involved in flood risk issues:

●   the absence of specific guidance from the Scottish Office on the consideration of flood related issues in the planning system;

●   the lack of formal mechanisms for preparing reliable flood risk maps which could be taken into account in preparing development plan policies;

●   the limited coordination between flood defence interests (with responsibilities split between riparian owners in rural areas and Regional Councils in urban areas) and land use planning interests.

Planning and Dam Safety

The Reservoirs Act 1975 provides for the safe design and maintenance of reservoirs in Great Britain, to prevent an escape of water that could devastate downstream areas. Currently, large raised reservoirs (with a capacity of over 25,000m^3 above the level of the adjacent natural ground) are placed in four hazard categories, mainly according to the level of downstream development (Table 9.3). If development is permitted downstream of Category B–D dams, then the **dam owner** could be subject to significant cost to improve the dam safety to withstand a more extreme flood event. For example, development on a floodplain previously used only for agriculture or recreation would require a dam upgrade from Category C to Category A with a design improvement from the 150 year flood to the 10,000 year flood. This could involve raising an embankment dam or enlarging a spillway at considerable cost.

It is clear, therefore, that development on floodplains or close to streams in rural areas could have significant implications for reservoir owners. In this context the planning system can have an important role in safeguarding private investment by diverting development away from undeveloped

Table 9.3 Reservoir flood and wave standards by dam category (after Ice, 1978).

CATEGORY	INITIAL RESERVOIR CONDITION	DAM DESIGN FLOOD INFLOW		
		GENERAL STANDARD	MINIMUM STANDARD IF RARE OVERTOPPING IS TOLERABLE	ALTERNATIVE STANDARD IF ECONOMIC STUDY IS WARRANTED
A. Reservoirs where a breach will endanger lives in a community	Spilling long term average daily inflow	Probable Maximum Flood (PMF)	0.5 PMF or 10,000 year flood (take larger)	Not applicable
B. Reservoirs where a breach (i) may endanger lives not in a community (ii) will result in extensive damage	Just full (i.e. no spill)	0.5 PMF or 10,000 year flood (take larger)	0.3 PMF or 1000 year flood (take larger)	Flood with probability that minimises spillway plus damage costs; inflow not to be less than minimum standard but may exceed general standard
C. Reservoirs where a breach will post negligible risk to life and cause limited damage	Just full (i.e. no spill)	0.3 PMF or 1000 year flood (take larger)	0.2 PMF or 150 year flood (take larger)	
D. Special cases where no loss of life can be foreseen as a result of a breach and very limited additional flood damage will be caused	Spilling long term average daily inflow	0.2 PMF or 150 year flood	Not applicable	Not applicable

areas at risk from dam failure which, at present, would cause only minor inconvenience to landowners.

Planning and Compensation

Although compensation is not available where planning permission is refused on the grounds of land instability or flooding (Town and Country Planning Act 1990 S.121 (7)), local planning authorities have a range of positive powers that may be used to control the use of land in vulnerable areas. Where it is in the interests of the proper planning of their area (including present and future amenity interests) a planning authority may, with the approval of the Secretary of State, serve a **discontinuance order** (under S.102 of the 1990 Act) to:

- discontinue any use of land;

- impose conditions on the continued use of land;

- require alterations to any buildings;

- require the removal of any buildings.

If the use of a discontinuance order results in depreciation in land value or the loss of enjoyment of the land, the planning authority is required to pay compensation to the affected parties, under S.115 of the 1990 Act, for:

- loss in land value;

- disturbance and expenses, including costs incurred in moving and loss of trade or business;

- rehousing, with the local planning authority obliged to provide alternative accommodation.

Property owners may seek to have the local authority purchase the land covered by a discontinuance order or where planning permission has been refused or revoked (S.137), provided the land is left without any reasonable beneficial use.

Under S.226 of the 1990 Act, a local authority may purchase compulsorily any land in their area which is required to achieve planning objectives within the area. The Secretary of State for the

108

Environment has similar powers for compulsory purchase which can be used to meet the interests of proper planning in an area, or to secure the best or most economic use of land (S.228). The acquisition procedure is controlled by the Acquisition of Land Act 1981 and the Compulsory Purchase of Land Regulations 1990.

These compensation and compulsory purchase powers **could** be used as part of a policy of **managed retreat** from, for example, unprotected eroding cliffs or areas where the replacement or improvement of existing defences may not be cost effective or environmentally acceptable. However, to the consultants' knowledge these powers have not been used for this purpose by local planning authorities; lack of resources and a reluctance to consider abandoning part or parts of existing communities are amongst the reasons why this approach is not considered a viable option in many areas.

Relocation of communities away from flood risk areas is, however, practised in the USA. Following the 1993 floods, the community of Chelsea, Iowa have decided to move to higher ground, half a mile away from the Otter Creek floodplain. The Federal Government will contribute over £4.5M in the relocation exercise, paying for the new infrastructure and providing around £7,500 to each homeowner to move their homes or build new ones. The decision to relocate was influenced by the fact that residents would have had to pay around £6,500 to raise the foundations of their homes before they could get flood insurance. However, the average house value had fallen to £2,500 after the floods.

Managed Retreat

The threat of rising sea levels has heightened awareness that it may not be possible to sustain the standards of protection provided by existing coastal defences. In addition, the decline in national value of agricultural output during the 1980s has reduced the economic benefit for protecting some agricultural areas. Managed retreat (also termed managed set-back or managed realignment), involving the abandonment or relocation of defences has been widely advocated as a coastal management option, especially where the policy could lead to significant conservation benefits (e.g. Posford Duvivier Environment 1991, English Nature 1992; Brooke, 1992). By progressively shifting the line of coastal defences inland new

intertidal areas may be created which, in turn, absorb wave energy and may reduce the need for high specification engineered defences. The newly-created intertidal areas could also offset losses of habitat to erosion or sea-level rise (English Nature, 1992).

Managed retreat should not be considered as a "do nothing" option. As MAFF (1993b) noted, the design of an effective scheme requires the development of a natural inter-tidal profile capable of absorbing wave energy and supporting valuable habitats. Indeed, the type of saltmarsh vegetation needs to be carefully assessed in the context of an understanding of the tidal regime and availability of littoral sediments.

Although the policy of managed retreat has gained official recognition – it is mentioned in both DoE Circular 30/92 (Development and Flood Risk) and PPG20 (Coastal Planning) – doubts remain over how such schemes may be implemented. The NRA, in England and Wales, can generally abandon a line of defence when it reaches the end of its residual life without becoming liable to compensation, with the exception of those areas where it has taken on the liability for defences from private landowners (i.e. **commuted liability**). However, there may be a need for compensation if they undertake habitat creation work that reduces the residual life of the defences and, hence, the value of private land. In such circumstances the NRA may purchase the land compulsorily or consider negotiating a management agreement with the landowner (under the Water Resources Act 1991).

However, a major obstacle to be overcome is the need to balance the strategic benefits with the losses faced by individual property owners. Posford Duvivier Environment (1991) observed:

"The politics of the retreat option cannot be ignored. History has demonstrated that British landowning interests are a politically powerful lobby, being both vociferous and effective in achieving their aims. Support for the principle and objectives of the retreat option ... is unlikely to be forthcoming in the absence of an adequate compensation provision."

For their part MAFF recently established a new **habitat creation scheme** (saltmarsh option) in 1993 which allows for twenty year set aside payments of between £195/ha (permanent pasture) and £525/ha (arable land) to encourage farmers to return intertidal land to saltmarsh.

Public resistance to retreat should not be underestimated. The responsibilities of coastal defence authorities are often mistakenly seen as mandatory rather than permissive. The apparent sacrifice of land and property for economic or conservation reasons is viewed as a failure in what is often regarded as a war against the sea. In this sense public perception is similar to the prevalent view of coastal engineers of over 50 years ago. However, whereas coastal defence strategies have evolved with improved understanding of physical processes and changing attitudes to the environment, public awareness has not moved on. Clearly, there is a need for informing the public and developers on this issue.

Summary

Avoidance of vulnerable areas can be an effective approach to reducing the impact of potentially damaging events. This approach may also help avoid public expenditure on flood and coastal defences or private investment in, for example, dam safety improvements upstream of new developments. However, it is not without its consequences. Decisions not to invest in existing property in vulnerable areas can lead to a loss in financial confidence and cause decline and decay. Restrictions on the siting of new development, through the operation of the planning system, effectively means increasing the pressure for development elsewhere within a region, county or district. In areas where available land is scarce this can lead to conflict between the need for hazard management and conservation objectives.

Local planning authorities can also have a positive role in ensuring that vulnerable areas are avoided, through the compulsory purchase of land or the use of discontinuance orders. These powers could be used to support a shoreline management policy of managed retreat in areas where improvement or replacement of existing defences may not be economically viable. However, the public's attitude that coastal defence should be provided by the government and the absence of a clear commitment to the compensation of coastal zone landowners are significant constraints to the effectiveness of managed retreat policies. If such policies are to prove successful there needs to be a programme of informing the public and developers.

It is clear that the prime responsibility for protecting land rests with the landowner. There appear to be, however, mechanisms for using

compensation payments for achieving coastal planning, shoreline management and conservation objectives. At present, however, there has been no government guidance on the use of such powers to achieve coastal management objectives, with consideration of the compensation issue absent from PPG20 Coastal Planning (DoE, 1992a), Circular 30/92 Development and Flood Risk (DoE, 1992b), and the MAFF/WO Strategy for Flood and Coastal Defence (MAFF, 1993a).

Chapter 9: References

Brooke J.S. 1992. Coastal defence: the retreat option. Journal of the Institution of Water and Environmental Management, 6, 151–157.
Department of the Environment 1992a. PPG20 Coastal Planning. HMSO.
Department of the Environment 1992b. Development and flood risk. Circular 30/92. (MAFF Circular FD 1/92; Welsh Office Circular 68/92). HMSO.
English Nature 1992. Campaign for a Living Coast. English Nature, Peterborough.
Highland Regional Council 1993. Flooding and Development. Development Plan Register, 4, 8.
Institution of Civil Engineers 1978. Floods and reservoir safety. Thomas Telford.
MAFF/Welsh Office 1993a. A strategy for flood and coastal defence in England and Wales. MAFF Publications.
MAFF 1993b. Coastal defence and the environment. MAFF Publications.
Posford Duvivier Environment 1991. Environmental opportunities in low lying coastal areas under a scenario of climatic change. Report to NRA, DoE, NCC and Countryside Commission.

10 Reducing the Likelihood of Damaging Events

Introduction

Although erosion, deposition and flooding are natural processes, land use and development can transform the hazard character of an area, modifying the intensity and frequency of many of these processes. It follows, therefore, that effective management of hillslopes, erosion, run off from new development, floodplains, river channels, coastal cliffs and the foreshore can contribute to reducing the size and frequency of damaging events and by ensuring that the effects of human activity on the occurrence of hazard–triggering conditions are minimised. In this context, many of the maintenance and repair activities described in Chapter 8 are of direct relevance, especially the prevention of obstructions to river flow or dredging to maintain an adequate channel capacity.

Hillslope Management

Although few farmers or foresters see soil erosion as a threat to their livelihoods in terms of lost yields (Armstrong et al, 1990), the supply of sediment to drainage channels, streams and rivers can create water storage and flood problems or navigation difficulties downstream. In some areas, erosion can also create considerable off–farm problems from mudfloods and deposition; such an event at Rottingdean in 1987 caused over £600,000 of damage. The rate of sediment delivery from hillslopes to watercourses is, however, likely to be only a small fraction of the estimated values of between 1 and 5 tonnes per hectare annual soil loss in erosion prone areas (Evans, 1988; Boardman, 1990).

Soil conservation measures have not been widely practised in Great Britain, mainly because they are not seen to be cost–effective, in the short term, for the landowner. However, it can be argued that the main justification for soil conservation may be to limit the potential nuisance created off farm by erosion events (see, for example, the Breaky Bottom litigation case described in Chapter 8). Grassing of valley floors, the creation of riparian barriers and the growing of winter cover crops could prove the most suitable measures (Table 10.1; Morgan, 1992).

Current hillslope management practices are, however, piecemeal solutions frequently applied on a farm–specific basis rather than for strategic control of soil erosion over broad areas. At present, the costs of such measures are borne by the farmer or landowner; this is seen by some to be a disincentive to more effective use of such measures. As Morgan (1992) wrote;

> "If soil conservation measures are to be effective, there must be incentives to encourage their adoption by farmers. These can include regulatory instruments, advisory work and financial support. They may need to be backed up by changes in overall agricultural policy because counter-incentives promoting agricultural practices that fail to protect the land against erosion would render soil conservation programmes pointless."(Morgan, 1992).

Financial incentives could spread the costs of erosion control across the community who, to a large extent, are the beneficiaries of soil conservation measures. In this context, a variety of **river channel–side zonation** policies (Newson, 1992) may provide financial incentives for controlling the supply of run–off and sediment into water courses. These include:

- the EC **Set Aside** policy which could be used to provide financial incentives for grassing valley floors and establishing riparian barriers;

Table 10.1 A range of soil conservation options (after Morgan, 1992).

APPROACH	COMMENT
Structural Measures	Channel terraces and bench terraces; costing up to £1000 per hectare, they are unlikely to be used in Britain.
Agronomic Measures	Grass leys can be used in a crop rotation to decrease erosion. Winter cover crops can prevent erosion of bare land prior to a spring-sown crop. They can also decrease nitrate leaching; their use is recommended by MAFF (1991) and SOAFD (1991) in their Codes of Good Agricultural Practice for the protection of water resources from pollution. Winter Sown cereals can be planted early enough to ensure adequate ground cover during the winter high erosion risk period.
Soil Management	Direct drilling of seeds without conventional tillage; this is considered unsuitable on silty and sandy soils (i.e. the main erosion prone soil types) because of the effect on yields. Contour cultivation; ploughing around the contours of a slope rather than across them. However, field layout often makes contour ploughing difficult, and steep slopes may make the practice unsafe.
Off-Farm Damage Control	Silt traps or emergency earth dams have been used in the South Downs. Grassing valley floors could decrease runoff velocity and induce deposition, preventing it from reaching watercourses. Creating strips of uncultivated land as barriers along riverbanks (riparian barriers).

- Countryside Stewardship schemes;

- the Environmentally Sensitive Areas scheme which includes river-related areas such as the Suffolk Valleys and the Test Valley.

Runoff Management

New development can significantly increase the quantity and rate of runoff reaching watercourses through the creation of extensive areas of impermeable materials. These effects can lead to flooding when the additional water causes the channel capacity to be exceeded during rainfall or snowmelt events. Problems are often associated with culverts, bridges and other channel constrictions.

The planning system is the principal mechanism for ensuring that development does not increase the risk of flooding elsewhere due to the generation of additional runoff (see Table 10.2 for a range of relevant planning policies). In England and Wales, the close relationship between local planning authorities and the NRA, as detailed in DOE Circular 30/92 (DoE 1992b;), is intended to ensure

that potential runoff problems are considered throughout the planning process, from development plan preparation to the determination of planning applications. The potential influence of the NRA is clear:

- it is a statutory consultee in the preparation of development plans;

- it is consulted on individual planning applications when significant flood defence considerations arise.

To encourage an effective and uniform approach to tackling surface water runoff issues, the NRA has prepared **guidance notes** (NRA, 1994; Table 10.3) advising local planning authorities to normally resist development which would result in an increase in the risk of flooding. Where appropriate, the guidance could be substantially replicated in development plans; if not, the NRA may formally object to the plans if it conflicts with their stated objectives.

During the development control process, the NRA may provide a broad assessment of the potential for increased flood risk and the scope of alleviation works that may be needed. It is the developer's responsibility, however, to investigate and design

Table 10.2 Examples of planning policies for runoff management, in England and Wales.

POLICY TYPE	COMMENT
Strategic	Berkshire CC; proposals would normally be resisted which could result in an increased flood risk due to additional surface water runoff.
Detailed	Brent BC; ensure that developments do not overload storm drains or watercourses.
	Bromley LBC; not normally permit development which increases flood risk due to additional discharge.
	Chiltern DC; not normally permit development which would increase the risk downstream due to additional surface water run–off.
	Harrogate BC; not normally permitted which generates surface water runoff likely to result in adverse impacts such as increased flood risk, river channel instability or damage to habitats.
	Leicester City C; storm water detention areas required in specified locations.

the alleviation works and demonstrate to the local planning authority how the flood risk will be controlled. The range of possible works include:

- provision of surface water storage areas such as temporary storage areas or ponds;

- flow limiting devices;

- infiltration through soakaway's (but see coastal cliff management, below).

Circular 30/92 (DOE, 1992b) advises local planning authorities that they should take steps to ensure that development is not brought into use until the necessary works have been completed, by:

(a) by **seeking the applicant's agreement** to the application being held in abeyance while he tries to make suitable arrangements;

(b) by **refusing permission** and giving advice on the kind of revised application which might overcome the difficulty;

(c) by seeking to impose a **negative** (or "Grampian" type) condition (see Circular 1/85);

(d) by entering into a **planning obligation** under S.106 of the Town and Country Planning Act 1990 (in such cases, the authority will need a formal agreement between the applicant, the NRA and the owners of the land through which the water would run, providing for the financing and carrying out of the necessary works and for

their future maintenance).

In Scotland, there is no equivalent to the NRA, although planning authorities may seek advice from the various Regional Council Drainage Departments (or the equivalent) who are a statutory consultee on sewerage issues. The technical approaches for dealing with runoff problems associated with new development are, of course, similar to those described for England and Wales.

Floodplain Management

The storage and infiltration of floodwater's on the low lying land adjacent to a river is an effective means of regulating the potential impact of a flood event. Reduction in **floodplain storage**, through the encroachment of development or the construction of flood embankments can lead to larger floods downstream, and may increase the risks to urban communities.

The planning system can be used to restrict floodplain development (see Chapter 9) and thereby prevent further reduction in floodplain storage. As was described above, for runoff management, the relationship between the local planning authority and the NRA is central to floodplain management in England and Wales (see Table 10.4 for the NRA's guidance note on protection of the floodplain). A range of typical planning policies for preventing a reduction in floodplain storage are shown in Table 10.5.

This restrictive approach to floodplain management can, however, lead to conflicts with minerals

113

Table 10.3 NRA guidance note for surface water runoff (from NRA, 1994).

Local/District Concerns	
Surface Water Runoff	
THE ISSUE	
3.5 Unless carefully sited and designed, new development or redevelopment can increase the rate and volume of surface water runoff. This can result in two types of problem. The first is the increased risk of flooding in areas downstream from the development in question. The second is physical damage to the river environment. This is a catchment wide issue and the NRA will take a co-ordinated approach to all developments.	
GUIDANCE STATEMENT	JUSTIFICATION
L6 The LPA should normally resist development which would result in adverse impact on the water environment due to additional surface water runoff. Development which could increase the risk of flooding must include appropriate alleviation or mitigation measures, defined by the LPA in consultation with the NRA and funded by the developer. Developers will be expected to cover the costs of assessing surface water drainage impacts and of any appropriate mitigation works, including their long-term monitoring and management.	3.6 Guidance for Planning Authorities on dealing with runoff from development is contained in DoE Circular 30/92 "Development & Flood Risk" (WO 68/92). New developments may result in a substantial increase in surface water runoff as permeable surfaces are replaced by impermeable surfaces such as roofs and paving. This may result in an increase in the risk of flooding downstream to an unacceptable level and a reduction in infiltration to groundwater. Other consequential effects include increased pollution, silt deposition, damage to watercourse habitats and river channel instability, as well as reduction in both river base flows and aquifer recharge. These effects can often be at some considerable distance from the new development. The LPA, in consultation with the NRA, will assess the surface water runoff implications of new development proposals. New developments will only be permitted where the LPA is satisfied that suitable measures, designed to mitigate the adverse impact of surface water runoff, are included as an integral part of the development. Where appropriate, the development should include provision for the long term monitoring and management of these measures. Arrangements under S.106 of the Town and Country Planning Act 1990 may be appropriate.

Table 10.4 NRA guidance note for flood plain protection (from NRA, 1994).

Local/District Concerns	
Protection of the Floodplains	
THE ISSUE	
3.3 Throughout England and Wales, and particularly in urban areas, a considerable amount of development has taken place on the coastal plain as well as in river floodplains. Consequently, people and property in these areas are already at risk from flooding. New development in floodplains is also likely to be at risk from flooding. The NRA holds information identifying many of the areas known to be at risk and will provide such information as required. Development can also have the effect of increasing the risk of flooding elsewhere.	
GUIDANCE STATEMENT	JUSTIFICATION
L5 Within the identified floodplain or in the areas at unacceptable risk from flooding the LPA should resist new development, the intensification of existing development or land raising. Where it is decided that development in such areas should be permitted for social or economic reasons, then appropriate flood protection and mitigation measures, including measures to restore floodplain or provide adequate storage, will be required to compensate for the impact of development. At sites suspected of being at unacceptable risk from flooding but for which adequate flood risk information is unavailable, developers will be required to carry out detailed technical investigations to evaluate the extent of the risk. In all cases, developers will be required to identify, implement and cover the costs of any necessary measures. In some cases the elements of the necessary measures may be such that they are best undertaken by the NRA itself, but in these cases the cost would be covered by the potential developers.	3.4 Guidance for Planning Authorities on protection of the floodplain is contained in DoE Circular 30/92 "Development & Flood Risk" (WO 68/92) and guidance on coastal floodplains is contained in PPG20 "Coastal Planning". In addition to the risk of flooding to the proposed development itself, development in such locations may increase the risk of flooding elsewhere by reducing the storage capacity of the floodplain, and/or impeding the flow of flood water. Land raising in the floodplain may have a similar effect. Consequently, the NRA looks to the LPA to resist development in such locations, while redevelopment of existing sites will only be considered where the LPA, in consultation with the NRA, is satisfied that the developer will provide appropriate mitigation and/or protection measures. There may also be opportunities to enhance or restore the natural floodplain when redevelopment takes place.

Table 10.5 Examples of planning policies for preventing a reduction of floodplain storage, in England and Wales.

POLICY TYPE	COMMENT
Strategic Policies	Cumbria CC; development not normally permitted where it would increase the risk of flooding elsewhere.
	Kent CC; development not normally permitted if it would be likely to increase the risk of flooding elsewhere.
	Mid Glamorgan CC; development likely to increase the risk of flooding will not be permitted.
	Suffolk CC; development will not be acceptable if it would impede materially the flow or storage of floodwater, increase the risk of flooding elsewhere.
Detailed Policies	Chiltern DC; not normally permit new development or the raising of land levels in areas at risk from flooding.
	Elmbridge DC; presumption against new development which would impede the flow of floodwater or reduce the capacity of the available washland.
	Newbury DC; proposals will be resisted if they inhibit the capacity of the floodplain to store flood water.
	Oxford City C; to prevent further serious risk of flooding elsewhere, no further raising of floodable land will be permitted on the Thames and Cherwell floodplains.

planning issues. Oxfordshire CC Minerals Plan, for example, includes areas of future gravel extraction in the floodplain. The NRA would not necessarily object to proposals in these areas provided the floodplain capacity was not reduced. This can restrict the restoration options for potential sites; whilst water uses may be satisfactory (and may increase the storage capacity of the floodplain), filling for restoration for agriculture could not involve "doming" to improve drainage conditions and achieve a good standard of re-instatement (see DoE, 1989; MPG7: The Reclamation of Mineral Workings). Stockpiles may also need to be kept to a minimum, with storage bunds or screens designed to prevent the obstruction of flood flows.

River Channel Management

Obstructions to flow or inappropriate river channel and bankside works can lead to an increase in flooding problems, either at the site or elsewhere downstream. In England and Wales, potential problems can be prevented by the need for consent from the NRA for any structure in, over or under a main river (the Water Resources Act 1991 S.109), works which might obstruct flow in a watercourse (outside an IDB area; the Land Drainage Act 1991 S.23) and compliance with byelaws made by the NRA under the Water Resources Act 1991 S.120. Although specific byelaws vary between regions, Table 10.6 highlights the range of general topics

that are usually covered. These can involve:

- serving notice on landowners or property owners to undertake maintenance or repair works;

- requiring the NRA's consent before proceeding with works within the river channel or a specified distance from the riverbanks (see Table 10.6);

- preventing certain activities such as excavating close to flood defences and placing the material on the riverbanks.

In general, NRA byelaws only concern **main rivers**. However local authorities and IDB's may make byelaws to secure the efficient working of flood defences on ordinary watercourses, under the Land Drainage Act 1991 S.66 (IDB byelaws apply to their whole area). Local authority byelaws can only be used to prevent flooding or mitigation of any damage caused by flooding. In Scotland, byelaws to prevent the obstruction of any watercourse and regulate the deposition of material on riverbanks can be made by the Regional Council under the Flood Prevention (Scotland) Act 1961 S.6. However, these can only apply to river channels in non-agricultural areas; and no byelaws are believed to be in existence.

Table 10.6 NRA land drainage byelaws: A selection of typical byelaws for river channel management.

TYPE OF BYELAW	SPECIFIC REQUIREMENT
Prohibition	• prevention of interference with river control works • obstruction to flow e.g. fall of vegetation or debris into the river • driving of animals or vehicles on riverbanks • damage to riverbanks by grazing • deposition on riverbanks
Consent	• alteration of the level or direction of flow • acts endangering stability of banks or defence works • heaps or construction on floodplains • planting of trees • dredging operations • construction close to riverside or flood defences Note: Distances from riverbanks for controlling construction vary between Regions:Anglian – 9m; Northumbria – 5m; North West – 8m; Severn Trent – 8m; Southern – 8m; South West – 7m; Thames – 8m; Welsh – 7m; Wessex – 8m; Yorkshire – 8m
Serve Notice	• maintenance and repair of river control works • control of vermin on banks • removal of vegetation from riverbanks • repairs to buildings etc.

Coastal Cliff Management

Coastal cliffs can be very sensitive to the effects of land use and development. Indeed, many of the reported landslides on the developed coastline have been, at least in part, caused by human activity. Problems are frequently associated with:

- the artificial recharge of groundwater levels e.g. through leaking water pipes, sewers and soakaways;

- excavation of slopes causing a loss of passive support or unloading of the material upslope;

- disruption of sediment transport following the installation of coastal structures such as groynes and breakwaters.

The planning system has an important role in cliff management by ensuring that development is suitable and takes full account of the potential instability problems that it may generate. A common approach is to require the developer to submit a **stability report** with a planning application which determines the site conditions and identifies any remedial measures which may be required to overcome any problems (Table 10.7). Much can also be done by the local authorities and other coastal managers to reduce the likelihood of slope failure in developed areas by simple, pragmatic cliff management practices (Table 10.8). In many areas preventing water leakage is likely to be the most cost–effective approach; in Ventnor, Isle of Wight this approach is central to the **Undercliff Management Strategy** developed to tackle the area's ground movement problems (Lee & Moore, 1991).

Provisions for an authority to ensure that disruption of sediment transport did not have significant effects are included within the Coast Protection Act 1949 S.16–17. However, these powers were not widely used and disruption of sediment transport

Table 10.7 Examples of planning policies for cliff management in areas of coastal instability.

Maldon DC;	applications for development in unstable areas should normally be accompanied by a stability report or statement that the site is stable. Any report should indicate how instability will be overcome and show how development will not endanger people or buildings on adjoining land.
North Norfolk DC;	a presumption against development that may increase coastal erosion as a result in changes in surface water runoff.
Poole BC;	planning conditions require no soakaways within 400m of cliff tops.
Shepway DC;	planning permission in areas of Sandgate will not be granted until a soil survey clearly demonstrates that the site can be safely developed and that the proposed development will not have an adverse effect on the landslide as a whole.
Weymouth & Portland BC;	it is necessary to produce a stability report as part of a planning application.

around the coastline is believed to have led to increased cliff erosion, especially on parts of the south and east coasts of England. Nowadays the potential problems are widely recognised by the Government (see MAFF/Welsh Office, 1993) and coast protection authorities, with the formation of coastal defence groups (see Chapter 3; Figure 3.2) helping to ensure that coastal engineering works by one authority do not have adverse effects on the neighbouring authority's coastline.

The **Government View Procedure** for determining the suitability of marine aggregate extraction licences is an important mechanism for ensuring that dredging operations do not affect coastal cliff erosion rates. Applicants are required to demonstrate that the proposed extraction will have no adverse effects on coastal defences or the environment generally. Although the procedure is not statutory, the Crown Estate Commissioners will not issue licences without a favourable Government view. Orders made under the Coast Protection Act 1949 S.18 can be used by local authorities to prevent the excavation or removal of materials on the sea bed or shoreline, although such powers have not been widely used in the past (Rendel Geotechnics, 1993). The Government View Procedure is currently under review.

Table 10.8 Examples of possible approaches to cliff management.

APPROACH	COMMENT
Construction Activity	• avoid inappropriate cut and fill operations • avoid excavations during winter months • avoid the removal of vegetation from cliff faces, unless part of a comprehensive treatment programme • voluntary control of the opening of trenches by statutory undertakers
Water Leakage	• prevent or reduce leakage from water mains and service pipes • avoid leakage from sewers • collect surface water runoff; avoid the use of soakaways and other natural percolation methods • avoid the use of septic tanks; ensure that they are regularly emptied and do not leak
Foreshore Development	• ensure that sediment transport to cliff foot beaches is not disrupted • ensure that coastal defence works do not concentrate erosion on adjacent unprotected cliffs
Offshore Activity	• ensure that offshore dredging does not lead to increased erosion risk

117

Foreshore and Sand Dune Management

Beaches, mudflats, saltmarshes and sand dunes can provide important natural coastal defences against flooding, as well as often forming an integral part of engineered schemes. These landforms can, however, be extremely sensitive to changes in the coastal environment, especially those resulting from human interference. Particular problems can arise from:

- disruption of sediment supply by foreshore structures, such as groynes, or dredging operations;

- removal of sediment for the aggregate industry (eg beach and dune sand) or for brick making (eg clays from mudflats).

A range of mechanisms can be used to minimise the potential effects of human activity on the integrity of natural coastal defences, including:

(i) **the planning system;** planning permission is required for mineral operations above LWM, with local planning authorities required to prepare **minerals local plans** (a special form of development plan). The need for caution in minerals planning in the coastal zone was highlighted in PPG 20 Coastal Planning (DoE, 1992a) which notes how inappropriate operations can lead to an increase in flood risk elsewhere.

Current examples of specific minerals policies which address the links between extraction and flood risk are not common. Grampian Regional Council, for example, has established a presumption against extraction from coastal sand dunes and all rivers (Grampian Regional Council, 1990). However, many coastal authorities recognise the value of beaches and dunes for conservation, tourism and recreation purposes; they have developed policies which restrict minerals activity in potentially sensitive areas, although they have often been worded to protect nature conservation interests. For example, in Mid Glamorgan, the County Council have prepared a Structure Plan policy which defines a presumption against sand extraction along the entire foreshore and dune systems of their coastline.

(ii) **sea defence byelaws;** in England and Wales, the NRA have powers to make byelaws under the Water Resources Act 1991 to ensure the efficient operation of coastal defences. As for the land drainage byelaws described above these vary between regions. Table 10.9 presents a summary of the relevant byelaws from the Southern Region which are typical of those adopted in regions with significant flood problems. Of particular note is the ability to control the removal or disturbance of materials from the sea bed and foreshore from below MHWM to up to 200m inland of the defences, and the prevention of development between the seaward side of the defences and up to 15m inland of the landward edge of any sea defences (natural or artificial).

(iii) **Coastal defence groups;** as was described earlier in the context of coastal cliff management the potential for disruption of sediment transport by coastal defence works is now widely recognised, with coastal defence groups providing a forum for ensuring that these issues are considered by neighbouring authorities when preparing a strategic management plan for a particular stretch of coastline.

(iv) **the Government View Procedure** for determining the suitability of applications for marine aggregate extraction licences. Here, the potential effect of extraction on coastal erosion is taken into account in the decision making process.

(v) **S.18 orders made under the Coast Protection Act 1949** can be used to prohibit the excavation or removal of materials from the seabed or shoreline.

Summary

There are a wide range of opportunities for reducing the likelihood of damaging events through effective management of human activity in and around sensitive elements of a catchment or the coastline. The main approach is to minimise the effects of development on the operation of erosion, deposition and flooding processes by requiring consents from various regulatory authorities for particular activities. The need for environmental assessment, as defined by EC Directive

Table 10.9 Summary of sea defence byelaws: NRA Southern Region.

Repairs to Buildings on Sea Defences;

Control of Animals;

Prohibition of acts endangering Stability of or causing Damage to Sea Defences:

-    removal or disturbance of materials on the sea bed in the vicinity of any groyne or other works in the sea that form part of the sea defences;

-    removal or disturbance of material from any part of the Authority Area below MHWM to 200m inland from the sea defences;

-    excavation or work that may cause damage to the defences, on any land or cliff adjoining the sea defences;

-    disturbance of vegetation growing on sea defences.

Erections, Excavations etc. affecting Sea Defence:

-    construction or excavation between MLWM and 15m of the landward side of any sea defence.

Driving of Animals and Vehicles on Sea Defences;

Deposit on Sea Defences:

-    placement of any goods or materials within 15m of the landward side of any sea defences.

Prevention of Interference with Sea Defences;

Maintenance and Alterations of Floodgates etc.;

Control of Vessels;

Control of Vermin on Sea Defences;

Damage to Sea Defences by grazing of Animals.

85/337/EEC complements these arrangements, although it is only required for certain types of major development.

Hillslope management relies upon individual landowners undertaking soil conservation measures at their own expense. However, subsidies and financial incentives available under various schemes (eg. Set Aside) **could** encourage land management practices that are compatible with soil erosion control objectives.

The NRA's supervisory role on all flood defence matters in England and Wales is a key factor in the ability to achieve effective flood hazard management in both the river and coastal environments. Of particular significance are the land drainage consents, byelaws for main rivers that are prepared under the Water Resources Act 1991 and its role in ensuring that flood defence issues are taken into account in the planning process (see DoE Circular 30/92). Here, the NRA

has two key functions:

● as a statutory consultee in the preparation of development plans. To ensure that local planning authorities are aware of the opportunities to tackle flood defence issues in development plans, the NRA have prepared **guidance notes** (NRA, 1994).

● as a consultee on planning applications in flood risk areas.

Local authorities can also have a significant role in many aspects of managing the effects of human activity on the levels of risk. They can make byelaws that address river channel management issues on non main rivers and can take an active role in coastal cliff management through the establishment of codes of practice for building and water control. Aspects of cliff management can be achieved through the planning system and building regulations for new development, but the legacy of unsuitable development, water supply and sewerage arrangements is likely to be cause for concern on some coastal cliffs. Key elements of the strategy, for landslide management adapted in the Isle of Wight Undercliff (Lee & Moore, 1991) may have broader relevance. These aspects are:

• sewer renewal;

• first time sewerage in areas of septic tanks and cess pools;

• surveys and replacement of leaking water mains and service pipes;

• a presumption against soakaway drains, with properties connected to closed drainage systems, where possible;

• encouraging local residents to maintain their property, especially to ensure that ditches are kept clear, etc. (Figure 8.1).

It is clear that much can be achieved within the framework provided by the existing legislation and policy advice. However, effective management appears to be constrained by a lack of awareness of current powers and, in the past, limited cooperation between the various authorities with an interest in erosion, deposition and flooding issues. In this context, the establishment of coastal defence groups to consider strategic coastal defence issues through **Shoreline Management Plans** and the NRA's programme of preparing **Catchment Management Plans** should be seen as important developments. In Scotland, the absence of an equivalent to the NRA does not mean that the various approaches described here cannot be used to manage flood problems. However, they appear harder to be employed effectively, as was widely highlighted after the 1993 Tayside floods (see Rendel Geotechnics, 1995).

Chapter 10: References

Armstrong A.C., Davies D.B. and Castle D.A. 1990. Soil water management and the control of erosion on agricultural land. In J. Boardman, I.D.L. Foster and J.A. Dearing (eds) Soil erosion on agricultural land, 569–574. John Wiley and Sons.
Boardman J. 1990. Soil erosion on the South Downs: a review. In J. Boardman, I.D.L. Foster and J.A. Dearing (eds) Soil erosion on agricultural land, 87–105. John Wiley and Sons.

Department of the Environment 1989. MPG 7: The reclamation of mineral workings. HMSO.
Department of the environment 1992a. PPG 20 Coastal Planning. HMSO
Department of the Environment 1992b. Development and flood risk. Circular 30/92 (MAFF Circular FD 1/92; Welsh Office Circular 68/92). HMSO.
Evans R. 1988. Water erosion in England and Wales 1982–1984. Report for Soil Survey and Land Research Centre, Silsoe.
Grampian Regional Council 1990. Minerals policy review.
Lee E.M. and Moore R. 1991. Coastal landslip potential assessment, Ventnor, Isle of Wight. Report to the Department of the Environment.
MAFF 1991. Code of good agricultural practice for the protection of water. MAFF Publications.
MAFF/Welsh Office 1993. A strategy for flood and coastal defence in England and Wales. MAFF Publications.
Morgan R.P.C. 1992. Soil conservation options in the UK. Soil Use and Management, 8, 176–180.
Newson M.D. 1992. River conservation and catchment management – UK Perspectives. In P.Boon, P. Calow and G.E. Petts (eds) River conservation and management, 385–396. John Wiley and Sons.
NRA 1994. Guidance notes for local planning authorities on the methods of protecting the water environment through development plans. NRA, Bristol.
Rendel Geotechnics 1993. Coastal planning and management: a review. HMSO.
Rendel Geotechnics, 1995. Erosion, Deposition and Flooding in Great Britain. Methodology Report. Open File Report held at the DoE.
Scottish Office Agriculture and Fisheries Department 1991. Prevention of environmental pollution from agricultural activity. Code of Practice. Scottish Office.

11 Protection against Potentially Damaging Events

Introduction

The main priority of hazard management must be the safeguarding of lives and minimising the stress and disruption caused by damaging events. Although the responses described in Chapters 8–10 can help reduce losses to life and property, there are many instances where the levels of risk are unacceptable without specific protection measures. Such measures can include **early warning systems** that ensure that people at risk can be moved to safety before events occur, constructing **defence schemes** to reduce the frequency of damaging events and **modifying building styles** to minimise the impact when damaging events do occur and secure evacuation routes in respect of flooding.

Early Warning Systems

Warning systems enable the emergency services to alert people at risk from imminent events and make advance preparations to lessen the impact of the event (see Chapter 8). In Britain, such systems are widely used to forecast and forewarn of flooding, but are becoming more widely used to monitor potential ground movement problems on the coast. Their obvious benefit is that they can give individuals time to move to safety or be evacuated by the emergency services. The success of a warning system is not simply a function of its technical basis; in practice there is a wide range of **behavioural factors** which can affect effectiveness (Green et al 1983). Recipients of warnings may not know what to do. They may decide to ignore the warning through lack of confidence in the accuracy of the prediction. In England and Wales, the general assumption is that an individuals main response will be to reduce financial losses by, for example, moving furniture upstairs.

Flood warning systems are well established in England and Wales, relying on a combination of storm weather forecasts by the Meterological Office, river flow gauges and flood warnings issued by the NRA. For river floods, weather radar is used by most NRA regions to identify the intensity and distribution of rainfall. Regional flood forecasting centres receive data from a local radar station and from the national network of weather radar sites. This data is used to estimate rainfall totals falling in a catchment and, together with rain gauge readings, forms the basis for providing advance warnings of flooding events to the Police and Local Authorities (**the National Flood Warning Services**; Table 11.1).

On the coast, the **Storm Tide Warning Service (STWS)** established, after the 1953 floods, to provide warnings of high surge tides on the east coast (Table 11.2). The distribution of surges is heavily biased towards the winter months when Equinox Spring Tides occur, with little surge activity experienced in the summer months. The STWS, therefore operates on a seasonal basis with 24 hour a day manning from 1st September to the end of April, with a general eye kept on the situation over the summer. The STWS is operated by the Meteorological Office on behalf of MAFF, who are responsible for ensuring that adequate warning procedures exist.

Warnings for East coast surges are issued for any of the Five Divisions (Figure 11.1) where a flood risk is calculated to exist for a particular tide. Within each division a "reference port" has been nominated and the tidal level determined at which sea defences in the division may be overtopped. On the south and west coasts, the STWS have developed a mathematical surge forecasting model and issue messages to the NRA regions, drawing their attention to any potential flood conditions.

In Scotland, the River Purification Authorities have powers to operate flood warning services under the

Table 11.1 NRA national flood warning service.

PHASE	COMMENT
1. Flood Forecasting	NRA staff monitor weather, rainfall and tidal and river levels continuously so that they know where flooding may occur, determine the severity of the flood, and the extent of areas that may be affected.
2. Flood Warning	Flood warnings are issued to the public via the police. The NRA aims to issue warnings for key areas four hours in advance of a potential flood, where it is feasible to do so. The National Flood Warning Service uses three flood warning codes, each one associated with the type of area and the flood risk. These are briefly summarised below: **Yellow Warning** Agricultural land and minor roads are likely to be flooded, but flooding of property is not expected. **Amber Warning** Agricultural areas and isolated properties are likely to be flooded **Red Warning** Residential and commercial properties are likely to be flooded.
3. Emergency Response	The Police and Local Authorities are responsible for ensuring that flood warnings reach those in threatened areas and for assisting them in relief from actual flooding. NRA flood defence teams ensure that rivers and culverts are kept clear to minimise the impact of the flood.

Agriculture (Scotland) Act 1970; existing systems include the Tweed RPB Flood Monitoring Scheme established in 1986. The Board took the view that it was not viable to provide a flood warning scheme for each important catchment, but that it could **monitor** the development of flood conditions. Accordingly, a monitoring scheme was established, comprising a network of 17 flood **marker boards** located at strategic sites where they can be easily read at flood times, alarm equipment installed at several river gauging stations and at seven 'headwater' alarm sites (Figure 11.2; Table 11.3). Elsewhere in Scotland, the Forth RPB operates systems on the Water of Leith and the River Tyne, and the Clyde RPB has a scheme on the White Cart. Following coastal flooding incidents in Strathclyde in 1990/91 the Clyde RPB commissioned feasibility studies for establishing coastal flood warning service on the Clyde; at present a service has been established at two locations on the river.

Table 11.2 The east coast Storm Tide Warning Service.

Primary Warning	Warnings are issued to the Police and NRA who set up local flood rooms and make initial preparations. Local colour-coded warnings will be issued by the NRA (see Table 11.1) and public warnings may be issued by the police, where necessary.
First Stage Warning	An <u>alert</u> is issued to the Police and NRA 12 hours before the time of high water at the particular reference port.
Second Stage Warning	A message is issued to the Police and NRA between 4–7 hours before high water; this can be one of three forms: – "cancel" if the threat has decreased – "alert confirmed" if the level is expected to be close to, or just above, danger level – "danger" if the danger level is expected to be exceeded by more than 0.2m

Note: On the West and South Coasts, the Meteorological Office pass mathematically – forecast surge levels to the NRA who assess whether their coasts are at risk.

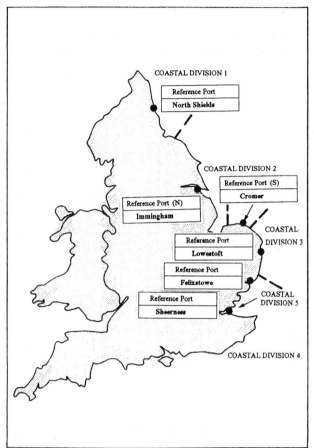

Figure 11.1 The east coast Storm Tide Warning Service: coastal divisions and reference ports.

consultants, in order to determine the most appropriate course of action (Figure 11.3).

Defence Schemes

Structural engineering measures represent the obvious, traditional and most publicly acceptable response to erosion deposition and flood hazards (Ward, 1978). The choice of option, or combination of options will be determined by the nature and scale of the problem, and the value of the property at risk. As emphasised in Chapter 2, the most severe threats to life and property are associated with flooding and coastal erosion, with a significant proportion of the population living in vulnerable areas. Although the responsibility for protecting property rests with the individual, legislation has empowered various authorities to undertake defence works that are in the **national interest**. These powers have been described in Chapter 3 and derive from:

- the Coast Protection Act 1949;

- the Water Resources Act 1991 (England and Wales);

- the Land Drainage Act 1991 (as amended by the Land Drainage Act 1994; England and Wales);

- the Flood Prevention (Scotland) Act 1961.

These Acts enable the relevant **operating authorities** to undertake defence measures and enable the government to offer financial support for specific works. As protection of life is the primary focus of Government flood and coastal defence policy (e.g. MAFF, 1993a), the order of priority for grant-aid is as follows:

- flood warning systems;
- urban coastal defence;
- urban flood defence;
- rural coastal defence and existing rural flood defence;
- new rural flood defence schemes.

These priorities are not prescriptive and grant-aid decisions are subject to rigorous appraisal procedures (MAFF, 1993b). In England and Wales, schemes (especially those that are grant-aided by MAFF or the Welsh Office) must be (Table 11.4):

In recent years there has been a marked increase in the use of early warning systems to monitor coastal landslide problems, especially where property or infrastructure are at risk. On the Isle of Wight, for example, a tiltmeter-based system has been in operation on the Military Road since 1981 to detect settlement behind the 70m high chalk cliffs (Barton & McInnes, 1988). The tiltmeters are connected to a central controller by means of an underground power supply. The system normally operates by triggering a warning when a pre-set amount of tilt is exceeded. The warning limits comprise an inner and outer limit representing a low and high level of hazard respectively. The inner limit activates an automatic telephone alarm to inform the Police in Newport. If the tilt continues to increase and the outer limits exceeded then internally illuminated traffic signs are activated at either end of the affected length of road. A similar system has been established in Ventnor, Isle of Wight, where a main road crosses an area of unstable land at the rear of the Undercliff. At this site a combination of automatic settlement cells and crack meters have been linked by modem to the local authority offices. In the event of the alarm being triggered, a procedure is in place for the local authority to inspect the site and seek advice from their

Figure 11.2 The River Tweed flood warning system.

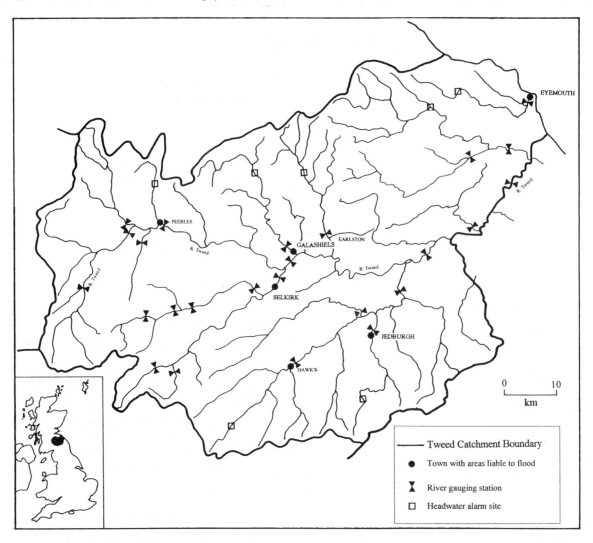

Table 11.3 The Tweed RPB flood monitoring scheme.

At each monitoring site, levels corresponding at ALERT, WARNING and DANGER conditions have been identified. At eleven gauging stations and at the headwater sites, solid state loggers are programmed to generate an alarm by telephone when the ALERT condition is reached. The programme routes alarm calls through a series of telephone numbers, commencing with the office number, followed by officers' home telephone numbers before finally calling the police station at Hawick, which is manned on a 24 hour basis. When an alarm call is received, contact with the Police is established immediately and monitoring commences by 'polling' the gauging stations by telephone, and by reading the flood marker boards. The police procedure is to warn members of the public when WARNING levels are exceeded at monitoring sites. Liaison is maintained with the Regional Council's Water and Drainage Department, and when there is a possibility of a major incident, a Flood Control Group is established, under the Regional Council's Flooding Contingency Plan, at regional headquarters. The Group is formed under the chairmanship of a senior police office, and comprises representatives of various Regional Council Departments (including Water and Drainage Services, Emergency Planning and Roads and Transportation), the Police, Fire Brigade and the Board.

In 1990, the Monitoring Scheme was extended to include five interrogable rain gauges. Generally the rain gauges are located adjacent to the alarm gauges so that they share the same logger and telephone lines. The location of the sites has been chosen so that an earlier assessment of a flood situation (from both rainfall and river level data) in the headwaters of selected catchments than is currently available can be made. In 1990, the flood marker boards were upgraded so that the ALERT, WARNING and DANGER levels could be identified by means of tape corresponding to a colour-code system. Three colours are used:

YELLOW (Alert)   –   A general alert or cautionary warning indicating the POSSIBILITY of flooding.
AMBER (Warning)   –   An indication that flooding is LIKELY in certain areas.
RED (Danger)   –   An indication that serious flooding is PROBABLE in specified locations.

Figure 11.3 The landslide monitoring strategy for the Ventnor Undercliff (after Rendel Geotechnics, 1995b).

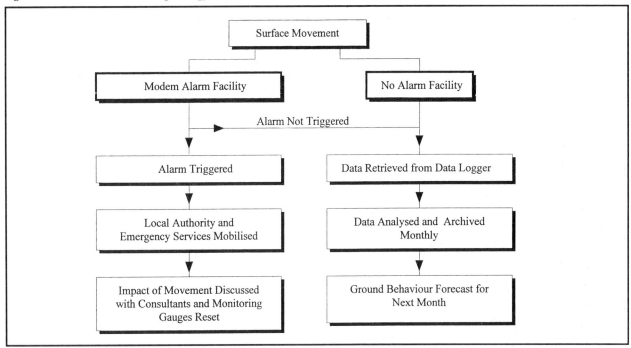

- **technically sound;**
- **environmentally acceptable;**
- **economically viable and cost–effective.**

Tables 11.5 and 11.6 provide a summary of the main forms of flood defence and coastal slope stabilisation measures. In general these defence measures can be sub–divided into two broad types:

- **passive measures** such as the provision of floodbanks, seawalls and flood alleviation channels;

- **active measures** such as flood gates and barriers, storage and diversion channels which can be operated in times of anticipated flooding.

It is important to stress that the construction of flood and coastal defence only **reduces** the risk. It cannot **eliminate** it. Much depends on the design life of the structure and the degree of risk that is acceptable. To be certain that flood defence works will not fail within ten years of completion they should be designed to cope with the 100 year event (Figure 11.4); such a structure would have a 45% probability of failing within 100 years. Thus, the tendency for increased investment and density of development behind defences may only lead to higher losses when, inevitably, floods occur that are larger than the ones which the defences were designed to cope with.

The **standard of protection** provided will, of course, vary with the nature of the land use in the area at risk. MAFF, for example, expresses **indicative standards** of **protection** in terms of the flood return period for five subjectively expressed current land use bands (Table 11.7). These standards are intended as guidance, not to set minimum standards.

In many instances, the standard of protection can give the general public a clearer indication of the hazard potential than the use of return period statistics. For example, communities on a floodplain may be protected by flood embankments that are designed to withstand a 1 in 25 year event. This standard may be consistent with that normally considered acceptable for rural villages, but may be unsuitable for more densely populated areas. Although the indicative standard of protection should not be regarded as an entitlement to a certain level of protection, it can highlight the relative risk associated with different land uses in an area.

The existing arrangements for flood and coastal defence have been very effective in protecting vulnerable communities. However, they have, in the past, tended to lead to a compartmentalised response to erosion, deposition and flooding processes whereas in reality their interaction is an important factor in initiating potentially damaging events. In the river environment, for example, channelisation works to prevent bank erosion can lead to channel adjustments both upstream and

Table 11.4 Criteria for appraising schemes in England and Wales (after MAFF/WO, 1993a).

Technical Soundness

- A range of options should be considered as part of the project appraisal, including a costed case for doing nothing, and in all appropriate cases, for example where an area immediately behind a defence is non-urban, whether setting back the existing line of defence could bring engineering and/or environmental advantages. This will ensure that full consideration is given to technical appropriateness as well as cost-effectiveness and environmental acceptability. However, there is no intention to promote any one particular option; solutions must suit the particular circumstances in each case.

- Schemes should be sustainable. This is they should take account of the interrelationships with other defences, developments and processes within a coastal cell or catchment area, and they should avoid as far as possible tying future generations into inflexible and expensive options for defence.

- Schemes should be based on an understanding of natural processes and, as far as possible, work with those processes.

Environmental Acceptability

- Grant-aid will be offered only for schemes which are judged environmentally acceptable. In all cases English Nature or the Countryside Council for Wales are consulted. The Countryside Commission, English Heritage or Cadw should be consulted where schemes may affect the character of the landscape or its recreational use, or the historic environment. In general schemes will not be approved if they are considered unsatisfactory by these bodies, although the Ministry or Welsh Office reserve the right to take their own view on the balance of interests in meeting the overall policy aim. Where the Environmental Assessment procedures lead to the preparation of an Environmental Statement that statement must accompany any application for grant-aid.

- The Ministry and Welsh Office will expect the potential impact on habitats and the environment generally to be a key consideration, and will start from the presumption that natural river and coastal processes should not be disrupted except where life or important man-made or natural assets are at risk.

- The cost of conservation and amenity works both for the reinstatement of an area to blend in with the existing environment and for works which further conservation objectives and enhance the environment, may be considered for grant-aid provided that the primary purpose of the works is flood defence or coastal protection.

- Where appropriate, schemes designed to manage water levels so as to achieve environmental benefits may be considered for grant-aid provided they meet the normal criteria of technical, environmental and economic soundness.

Economic Viability and Cost Effectiveness

- Schemes should have a benefit to cost ratio of at least unity to be considered for a grant. It is the Ministry's and Welsh Office's normal approach to seek to maximise the benefit to cost ratio from the options available.

---

downstream, creating increased flood risk or erosion problems for other landowners. The construction of coast protection works may lead to an increase in flood risk elsewhere as saltmarshes, mudflats, beaches and shingle ridges, which provide natural defences to low lying areas behind, are starved of sediment. However, in the last decade there have been significant changes in

Table 11.5 A summary of the range of approaches to flood defence.

Flood Regulation	• reservoirs for temporary storage of floodwater • barriers to control tidal surges
Flood Relief	• construction of bypass and diversion channels to carry some of the excess floodwater away from the protected area
Channel Improvements	• enlarging channel capacity, by straightening, widening or deepening
Flood Protection	• embankments and sea walls to confine the floodwaters
Beach Management Structures	• groynes, breakwaters and strong points to promote a natural defence • beach nourishment

Table 11.6 Principal method of coastal slope stabilisation.

APPROACH	METHODS
Excavation and Filling	– Remove and replace slipped material – Fill to load the slope base
Drainage	– Lead away surface water – Prevent build-up of water in tension cracks – Blanket the slope with free draining material – Installation of narrow trench drains aligned directly downslope, often by shallow drains laid in a herring bone pattern – Installation of interceptor drains above the crest of the slide or slope to intercept groundwater – Drilling of horizontal drains into a slope, on a slightly inclined gradient – Construction of drainage galleries or adits, from which supplementary borings can be made – Installation of vertical drains which drain by gravity through horizontal drains and adits, by siphoning or pumping
Restraining Structures	– Retaining walls founded beneath unstable ground – Installation of continuous or closely spaced piles, anchored sheet or bored pile walls – Soil and rock anchors, generally pre-stressed
Marine Erosion Control	– Control of toe erosion by crib walls, rip-rap, revetments, groynes – Control of surface erosion – Control of seepage erosion by placing inverted filters over the area of discharge or intercepting the seepage
Miscellaneous Methods	– Grouting to reduce ingress of groundwater into a slide – Chemical stabilisation by liming at the shear surface, by means of lime wells – Blasting to disrupt the shear surface and improve drainage – Bridging to carry a road over an active slide – Rock traps to protect against falling debris

Figure 11.4 The relationship between the standard of protection (as expressed by a return period event), the design life of a structure & the chance of failure during this period.

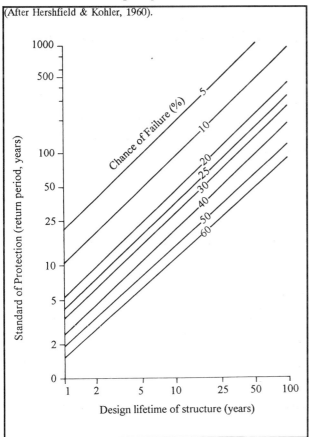

(After Hershfield & Kohler, 1960).

attitudes towards the use of engineering schemes to management flooding and coastal erosion problems. The main changes have been the recognition of a need for strategic planning of flood and coastal defence issues through **Catchment Management** and **Shoreline Management** (see Chapter 3), and the advocacy of soft engineering approaches on the coastline.

Traditional coastal defences have involved a combination of breakwaters and sea walls designed to oppose wave energy or flood embankments and barrages designed to reduce the threat of elevated water levels. These traditional measures are known as **hard engineering** solutions in which natural forces are opposed by structures. The success of this approach is clearly demonstrated by the extensive development which has taken place along the coast where, previously, natural hazards would have caused considerable damage. However, the presence of these fixed structures can generate problems for coastal managers. Most notably, these problems include the disruption of sediment transport and the reduction in intertidal area and, hence, coastal habitats (ie. "coastal squeeze"). There has, however, been a recent shift towards approaches which aim to work with natural systems by manipulating coastal processes to the benefit of

Table 11.7 Indicative standards of protection (after MAFF, 1993b).

CURRENT LAND USE	INDICATIVE STANDARDS OF PROTECTION (RETURN PERIOD IN YEARS)	
	TIDAL	NON–TIDAL
High density urban containing significant amount of both residential and non–residential property.	200	100
Medium density urban. Lower density than above, may also include some agricultural land.	150	75
Low density or rural communities with limited number of properties at risk. Highly productive agricultural land.	50	25
Generally arable farming with isolated properties. Medium productivity agricultural land.	20	10
Predominantly extensive grass with very few properties at risk. Low productivity agricultural land.	5	1

environmental interests as well as protecting coastal communities (MAFF, 1993c). Examples of this "**soft engineering**" approach include beach feeding and managed retreat. Beach feeding is the most widely used soft engineering technique and involves counteracting natural loss of beach sediment by artificial recharge (Brampton, 1992; West, 1992). The source of material is usually offshore (See Chapter 8).

Most flood and coastal defence works require planning permission from the local planning authority. The notable exception is flood defence improvements which are permitted development (under the GDO 1988 Part 15); as many flood defence schemes involve the replacement or enhancement of existing defences it is rare for planning permission to be needed. However, these permitted development rights can be withdrawn, although the local authority may be liable for compensation if permission is subsequently refused. It should not be assumed that proposed schemes will receive planning permission. Indeed, conflicts can arise between the operating authority and the planning authority over a variety of land use issues, especially landscape and conservation policies as will be illustrated by the following examples from the Maidenhead area and Christchurch Bay.

The Windsor, Maidenhead and Eton Flood Alleviation Scheme, for example, has recently highlighted the problems of reconciling flood defence interests with broader environmental issues (Rendel Geotechnics, 1995). In this instance Thames Region NRA had proposed a flood alleviation scheme comprising three main elements:

(i) upgrading the existing channel passing through the centre of Maidenhead;

(ii) improving the channel of the River Thames to ensure it is capable of carrying increased flows, with the construction of flood banks and walls;

(iii) construction of a new channel leaving the Thames at Maidenhead, running north of Dorney and Eton Wick close to the M4 and passing around the north and east sides of Eton College playing fields to rejoin the Thames downstream.

As part of the proposed scheme would involve the extraction of around 4.5m tonnes of sand and gravel from the new channel, the proposals were treated as a **mineral application** (for which outline planning applications are not allowed). Hence, the scheme required the consideration by the mineral planning authorities (Buckinghamshire County Council and Berkshire County Council) rather than the district councils. During consideration of the application both counties expressed various concerns about the scheme, especially the impact downstream and possible conflict with Green Belt and local landscape policies. In addition, the need for the scheme was questioned by residents in Buckinghamshire where there was strong local opposition to this proposal which was seen as being to provide flood relief (and so benefit) to properties in Berkshire, many of which have been built or purchased in the knowledge that flooding would be likely to occur, whilst the environmental impact of a large part of the scheme would fall in

Buckinghamshire. Buckinghamshire County Council also expressed the view that the NRA had not considered the proposals in the context of the Structure or Local Plans.

With the background of local opposition to the scheme, the MP for South Beaconsfield, Mr Tim Smith, secured an adjournment debate in the House of Commons (held on the 7 November 1991). Subsequently, the Secretary of State issued an Article 14 Direction and the application was called-in in February 1992. A planning inquiry was held 18 October to 17 December 1992. Scheme approval was granted in November 1994.

Similar problems were experienced in Christchurch Bay during 1989-1990, where a coast protection scheme proposed by Christchurch Borough Council generated objections from the then Nature Conservancy Council (NCC) who felt that the scheme would diminish the value of the site as a geological SSSI. On the basis of this objection the Secretary of State "called-in" the planning application. The site lies outside the Borough of Christchurch and, hence, permission was needed from the neighbouring local planning authority, New Forest District Council. New Forest D.C. informed the Secretary of State that they would refuse the proposal on the grounds of the possible effects on the SSSI, particularly in light of the policies set out in the relevant structure and local plans. Had the site been in their own area, Christchurch B.C. would have indicated approval; their subsequent lobbying of New Forest D.C. officers led to the decision being reversed. A public inquiry was still required to examine the NCC objections under the Coast Protection Act 1949 and as part of the "call-in" procedure for determining planning applications. Tyhurst (1991) notes that if the site had been within Christchurch B.C.'s area the conflict between coast protection and planning could have been resolved internally by the full Council rather than involving a public inquiry.

The following sections are intended to illustrate the variety of schemes that are currently being promoted by the different operating authorities. An indication of the scale of expenditure emphasises the importance that is placed in flood and coastal defence activity and also highlights the contrasts between the situation in England and Wales with that in Scotland.

NRA Flood Defence Expenditure

The total annual expenditure on flood defence is currently in the order of £240M (NRA, 1992); this is relatively evenly spread between sea, tidal and river defences, although there is a strong emphasis on works in urban areas (60 – 75% of expenditure). The bulk of the NRA's expenditure is split between **maintenance** and **improvements;** for the year 1992/93 over 40,000km of main river maintenance and 70km of river defence improvements were undertaken.

The capital expenditure programme concentrates on **cost effective urban schemes**. These can vary considerably in scope and cost, with current major spends over £2M identified on Figure 11.5. The largest scheme is the Maidenhead Flood Alleviation Scheme promoted by NRA Thames Region, valued at £74.3M.

Local Authority Flood Defence Expenditure in England and Wales

The costs of local authority promoted capital schemes are raised by:

(i)   **local authority funding arrangements;** flood defence expenditure is reflected in the **Standard Spending Assessments** for local authorities. This revenue support grant through credit approval and support grant covers almost all the costs not provided for by the **MAFF/WO** grants (see (ii) below).

(ii)   **MAFF/WO grant aid;** currently set at 25% with a 20% supplement for tidal and sea defence work. For grant-aided schemes in England, the Ministry issues **Supplementary Credit Approval** to enable local authorities to borrow in order to cover the residual costs. In Wales such approval is covered within a wider credit approval for local authorities. However, where flood defence schemes require greater levels of borrowing than can reasonably be afforded within these wider resources, they can be considered for **Special Project Approval.**

(iii)   **Other sources;** in addition to contributions made by landowners and developers, some local authorities are known to have

Figure 11.5 NRA major flood defence schemes 1992–1997 (NRA, 1992).

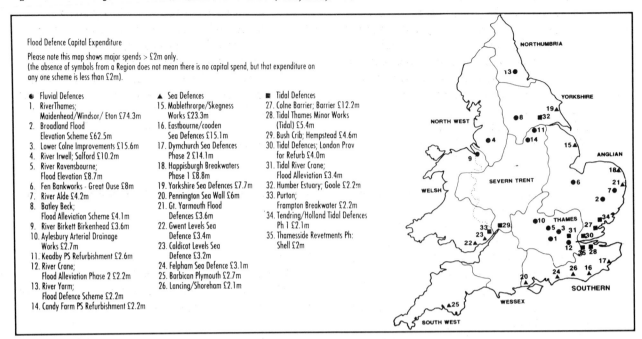

Flood Defence Capital Expenditure

Please note this map shows major spends > £2m only.
(the absence of symbols from a Region does not mean there is no capital spend, but that expenditure on any one scheme is less than £2m).

● Fluvial Defences
1. River Thames;
   Maidenhead/Windsor/ Eton £74.3m
2. Broadland Flood
   Elevation Scheme £62.5m
3. Lower Colne Improvements £15.6m
4. River Irwell; Salford £10.2m
5. River Ravensbourne;
   Flood Elevation £8.7m
6. Fen Bankworks - Great Ouse £8m
7. River Alde £4.2m
8. Batley Beck;
   Flood Alleviation Scheme £4.1m
9. River Birkett Birkenhead £3.6m
10. Aylesbury Arterial Drainage
    Works £2.7m
11. Keadby PS Refurbishment £2.6m
12. River Crane;
    Flood Alleviation Phase 2 £2.2m
13. River Yarm;
    Flood Defence Scheme £2.2m
14. Candy Farm PS Refurbishment £2.2m

▲ Sea Defences
15. Mablethorpe/Skegness
    Works £23.3m
16. Eastbourne/cooden
    Sea Defences £15.1m
17. Dymchurch Sea Defences
    Phase 2 £14.1m
18. Happisburgh Breakwaters
    Phase 1 £8.8m
19. Yorkshire Sea Defences £7.7m
20. Pennington Sea Wall £6m
21. Gt. Yarmouth Flood
    Defences £3.6m
22. Gwent Levels Sea
    Defence £3.4m
23. Caldicot Levels Sea
    Defence £3.2m
24. Felpham Sea Defence £3.1m
25. Barbican Plymouth £2.7m
26. Lancing/Shoreham £2.1m

■ Tidal Defences
27. Colne Barrier; Barrier £12.2m
28. Tidal Thames Minor Works
    (Tidal) £5.4m
29. Bush Crib; Hempstead £4.6m
30. Tidal Defences; London Prov
    for Refurb £4.0m
31. Tidal River Crane;
    Flood Alleviation £3.4m
32. Humber Estuary; Goole £2.2m
33. Purton;
    Frampton Breakwater £2.2m
34. Tendring/Holland Tidal Defences
    Ph 1 £2.1m
35. Thameside Revetments Ph:
    Shell £2m

---

obtained funds for flood defence–related works through the EC European Regional Development Fund (see below).

A brief review of local authority expenditure on flood defence reveals that there is no consistent pattern to the funding arrangements for schemes. Each authority appears to adopt a variety of strategies depending on their circumstances and the nature of the proposed works. Three examples are briefly described below to illustrate this complexity.

(i)   **City of Bradford MC;** the council have recently undertaken a major flood defence scheme for the Bradford Beck, a tributary of the River Aire, which flows through 21km of urban area. The beck is generally culverted, but suffers from severe flash flood problems. A flood on 6th June 1982, for example, resulted in an estimated £2.1M of damage, and 2 children drowned as they took shelter from the rain in an open channel section (Hutchinson 1992). Such flood events may occur every 10 years.

It has been reported that these flood problems have caused some loss in business confidence. Many businesses within the flood risk area found it difficult to obtain cost–effective insurance and developers were deterred by the additional costs imposed by the problems

(Hutchinson, 1992).

The City lies within an **Assisted Area** as part of an overall programme of urban regeneration, the council designated 13ha of land at risk from flooding as industrial and commercial areas. Major developments were planned in these and adjacent areas with a combined value of £250m. Flood defence, therefore, was considered an essential requirement of the regeneration programme.

The **Bradford Beck Flood Alleviation Scheme** comprises a 2.15km long diversion for storm flows, bypassing open channel reaches and inadequate culverts in the city centre. Around 1.8km of the scheme is being constructed in deep tunnel with associated shafts through carboniferous lower coal measures rocks; the remainder is in situ concrete construction, in an open–cut.

As the City lies within an **Assisted Area** the various economic regeneration projects were part funded by the EC as part of the council's approved **Integrated Development Operation Programme**. The Bradford Beck Scheme formed one element of the associated infrastructure subprogramme. The EC grant–aid contribution to the scheme was subject to economic appraisal by the DoE, based on an assessment of a cost/benefit analysis and a statement on the potential impact on economic regeneration; the grant amounted to £5.46M (around 50% of the total scheme cost). Grant–aid was approved by MAFF after a separate economic appraisal, covering 26% of the scheme cost after deduction of the EC grant. After a contribution of £0.85M from Yorkshire Water, the remainder of the costs were met by the City of Bradford M.C. (£3.99M).

(ii) **Tewkesbury BC;** the council have taken an active role in the resolution of drainage problems on non–main rivers and have undertaken a range of schemes with costs varying from £1,500 to £250,000, with a recent total expenditure of around £450,000. Small flood alleviation schemes are also undertaken as a rolling programme with the County Land Agent and the Highway Authority; the current annual expenditure on such works is around £35,000. The council have generally funded these schemes through their Capital Allocation and contributions from landowners and developers; they have not sought MAFF grant–aid.

Key problems generally relate to land drainage infrastructure in new housing developments. For example, in Cheltenham a housing policy area will provide 2,600 residential units, schools and community centres. The council is committed to £450,000 for the necessary infrastructure works. However, these costs will be recovered in full, including administration and financing costs, through multiple agreements under the Land Drainage Act 1991 S.20.

(iii) **Erewash BC;** although the southern area of the Borough lies within the River Derwent and Trent floodplains, there are a number of non–main tributaries which tend to have a very flashy response to rainfall and are affected by backing–up problems when the main–rivers are in spate. Following severe flooding of local streams in early 1977, a number of major schemes have been constructed forming the **Long Eaton and Beaston Flood Alleviation Scheme**, which was complete in 1992. The total cost of £3–3.5M was funded partly by grant–aid from MAFF and partly from local authority sources.

Small schemes are continuously being undertaken by the council, mainly to reduce the effects of development; some of which have been largely funded by private householders in affected areas (under the 1991 Act S.20). For example, the scheme at Risley Brook was completed in 1991 at a cost of £15,000, which was raised mainly from the affected householders.

Local Authority Flood Defence Expenditure in Scotland

The powers available to Regional Councils under the Flood Prevention (Scotland) Act may be broadly classified as relating to capital schemes and maintenance provisions:

- **Promotion of Flood Prevention Orders** – Such an order may be promoted by a Regional Drainage Authority in order to procure engineering measures to control flooding in non agricultural (predominantly urban) situations. The formal procedures require extensive consultations and notification of interested parties during the development phase of preparation of an order.

- **Operational Responsibilities** – Regional Drainage Authorities can, at their discretion, exercise powers under the Act for the routine maintenance, and repair of **urban watercourses** for the purpose of preventing or limiting the impact of flooding. An authority is only obliged to undertake maintenance work to watercourses where it is itself the riparian owner. This obligation extends to all food prevention schemes constructed within the authority area. Financial constraints are such that the approach of all authorities tends to be one of emergency maintenance in response to specific incidents of flooding, there being an obvious reluctance to set precedents by embarking upon extensive maintenance programmes and thereby assuming responsibilities which fall to many riparian owners.

The operation of the flood defence legislation is widely perceived by Scottish local authorities to be limited by the lack of opportunity for integrated management of flood problems throughout a catchment (Table 11.8).

Grant aid is available from the Scottish Office for flood prevention works under the Flood Prevention (Scotland) Act 1961; the grant is currently fixed at 50% which is available for capital works rather than management or maintenance. It is important to note that cost/benefit criteria **are not** considered by the Scottish Office in assessing whether a proposed scheme qualifies for grant–aid. Under the 1961 Act, a local authority may receive contributions towards the expenditure from any other person (S.12) and the Act does not preclude authorities from seeking additional sources of funding such as from the EC. If such circumstances arose, the Scottish Office would operate the principle of **additionality**, by which the 50% grant would be available only towards the residual costs, after deduction of the contributions from alternative sources of funding from the qualifying value of the scheme. However, it does not appear that Scottish local authorities have generally received significant contributions from potential alternative sources for financing their schemes.

Since 1961 there have been 63 flood prevention schemes confirmed under the 1961 Act of which 22 were undertaken prior to the establishment of the regional councils in 1975. Although actual cost information is difficult to obtain, it is clear that no scheme prior to 1975 cost as much as £1M and many were probably considerably lower. From 1975 to date a further 41 schemes have been implemented and 26 have been proposed. Although the largest scheme, currently under construction at Brockburn and Levern in Strathclyde, is estimated to cost £4M, the majority of the projected schemes are valued at far less that £1M, averaging £160k each.

In general, the limited number of such schemes can be attributed to:

• less widespread severe flood risk in Scotland than in many lowland areas of England and Wales;

• the protracted procedures by which flood prevention schemes receive grant–aid approval, which may include a Public Inquiry. For example, in Central Region, the time taken to obtain a flood prevention order at Bridge of Allen was reported to have been around 15 months for a £180,000 scheme, following several years

of ongoing discussions with the riparian owners.

Coast Protection

The sources of funding for local authority coast protection schemes are similar to those outlined above for England and Wales although county councils are also obliged to contribute to the cost of a scheme. At present, MAFF and the Welsh Office make available around £20M per year for coast protection schemes. There are numerous examples of coast protection schemes around the coast using a variety of techniques; the Whitby scheme completed in May 1990 clearly demonstrates the way in which a variety of schemes need to be considered with the final preferred scheme then subject to cost benefit analysis, (Clark & Guest, 1991).

The affected section of the Whitby coast involves 750m of cliffs on the northern flanks of the town which are composed of easily eroded glacial tills and fluvio–glacial gravels overlying Middle Jurassic sandstones, siltstones and mudstones. Marine erosion has caused repeated destruction, with about 50m of cliff top recession over the last 100 years so that increasing numbers of houses have come to be threatened.

Initial geomorphological investigations undertaken in 1983 identified the problem as due to combination of basal erosion by wave attack and high groundwater tables within the slope. Following an extensive ground investigation in 1985, a series of alternative engineering solutions were designed. On grounds of cost, safety, aesthetics and since the elevation of the works was fixed by the existing promenade the finally selected option was a rock armour wall together with slope reprofiling and drainage.

The finally adopted scheme was subjected to a cost benefit analysis as part of the scheme justification by the funding bodies (MAFF and the EC Regional Development Fund, Scarborough BC and North Yorkshire CC). The estimated cost of the construction, plus maintenance costs and consultant's fees were compared with the monetary value of the benefits derived from the project. The tangible benefits were the monetary losses that would be prevented by the scheme which are related to the protection of property, infrastructure and safeguarding the tourist industry.

Table 11.8 Flood defence in Scotland: A summary of the limitations in the existing legal framework.

The operation of the Flood Prevention (Scotland) Act 1961 is reported to be constrained by several factors which are considered to be a source of frustration to the Regional Drainage Authorities charged as the responsible bodies with the duty of administering and implementing the provisions of the Act; and which are not consistent with comprehensive management and engineering of flood risk within a river basin system:
• Powers limited to non agricultural flood risk such that up stream agricultural works conducted under the Land Drainage (Scotland) Act 1930 may proceed without regard for their impact on downstream conditions and may in certain circumstances render a Flood Prevention Scheme ineffective.
• There is a reluctance to embark upon schemes, particularly those of a minor nature, because of the complex administrative procedures to be observed.
• Powers under the Act are permissive, giving an Authority flexibility in its approach and implementation. Without a comprehensive flood register and a consistent system of reporting, the relative frequency and severity of flood incidents within an authority area may be subject to considerable subjective opinion in attempting to assess the relative impact of particular problems of flooding in order to prioritise actions.
• Perhaps the greatest single difficulty is one of public awareness and understanding of the apportionment of responsibilities in respect of flooding. Particularly in heavily populated urban centres the public appear generally unwilling to accept that their local authority is not responsible for the relief of flooding and the maintenance of watercourses.
• Most of the recurrent problems or urban flooding have been found to be localised and a consequence of inadequate maintenance of culverts and watercourses. In this respect the Act provides no remedy.

The potential losses were calculated on the value of property that would be lost over the immediate, short, medium and long term based on predictions of cliff top recession. Immediate was taken as 1 – 5 years, short term was based on a 5 – 15 year period, medium term on 15 – 65 years and long term on 65 – 100 years. Each property and infrastructure value was discounted to derive a present day value using the public sector test rate of 5% which was applicable at the time (1986). In addition to property losses, values were placed on the potential loss to the tourist industry of the area if no works were carried out and this was discounted over 50 years to derive a present day value. The ratio of the present value benefits to present value costs was 1.70. The 750m length of stabilisation eventually cost a total of £3M, of which one–third was spent on coastal defences and the remainder on earthworks.

Conflict with Conservation Interests

Despite the broad consultation requirements for flood and coastal defence schemes the promotion of engineering solutions can lead to conflicts of interest between local property owners and national conservation priorities, especially on the coast. Owners of vulnerable developments, such as properties in risk areas, naturally apply pressure on local authorities to provide defence measures to safeguard their investments. Such defences are, of course, funded by the public purse through MAFF grants and local authority contributions, with many schemes costing very large sums. Although cost/benefit analyses are undertaken to justify such expenditure the whole issue of coast defence is becoming increasingly politicised with local councillors and MPs often being drawn into debates over particular schemes as a result of lobbying by those whose house or livelihood is threatened. In such circumstances, the choice of the engineering response has often led to conflict with conservation agencies when, for example, nationally or internationally important habitats or geological sites are affected in order to protect property. An example is the recent debate over a coastal defence scheme proposed for Chewton Bunny, Christchurch described earlier (Tyhurst, 1991).

The balance between protecting vulnerable properties or communities and the need to preserve the character of the unspoilt coast also can lead to considerable conflict. For example, at Easton Bavents on the Suffolk coast, the Countryside Commission successfully opposed Waveney District Council's plans to protect a number of cliff top homes because of the possible effect of the proposed scheme on an Area of Outstanding Natural Beauty (McIlroy, 1990).

From the geological conservation perspective, the design of coast protection works should aim to control erosion sufficiently to protect the coast whilst allowing a small degree of erosion to maintain its exposure. For their part the former NCC commissioned " **A Guide to the Selection of Appropriate Coast Protection Works for Geological SSSIs**", which was undertaken by HR Wallingford (1991). This document considers appropriate engineering designs or alternative action, such as land purchase and compensation, for a variety of coastal environments (Leafe, 1991).

The need for environmental guidance in selecting appropriate coastal defence solutions was identified by MAFF in 1991. As a result they commissioned the Institute of Estuarine and Coastal Studies (Hull University) to prepare guidance to coastal engineers, planners, managers and environmental issues in coastal defence, the natural physical and ecological processes involved and the engineering techniques available (MAFF, 1993c). It embraces the principle of **multiple sustainable use** of the coast which attempts to achieve the solution of coastal management problems through the use of environmentally sensitive schemes involving the use of natural coastal systems. Similar guidance is currently in preparation for the river environment (Holmes, 1993) which would complement a number of existing handbooks, including:

- Code of Practice on Conservation, Access and Recreation (DoE/MAFF/WO, 1989);

- Conservation Guidelines for Drainage Authorities (MAFF/DoE/WO, 1991);

- Environmental Procedures for Inland Flood Defence Works; (MAFF/EN/NRA, 1992).

- IWEM Water Practice Manual (Brandon, 1988).

Compensation as a Flood and Coastal Defence Option

Under common law the primary responsibility for protecting land lies within the owner. Operating authorities are not required to protect land, although they have the power to do so when it is in the national interest. In such circumstances a range of scheme options will be considered, from "do-nothing" to varying degrees of protection, to find an acceptable balance between risk reduction, economics and environmental factors. Compensation is not available to landowners where the "do-nothing" option is favoured.

The current approach assumes that sites are valuable only in terms of the development infrastructure and services at risk and that the primary benefits are the damage and loss which schemes are expected to prevent or delay. However, cliff erosion and flooding can give rise to significant environmental benefits such as the supply of littoral sediment and the maintenance of conservation sites. In some instances these benefits can be viewed as being in the national interest. In such circumstances it might be appropriate to provide financial assistance to a landowner for maintaining the positive effects of the do nothing option, provided this can be demonstrated to be cost-effective. Such an approach would, of course, require a change in the current legislative framework but could contribute to easing the ever increasing conflict between the provision of defence works and environmental interests.

Modifications to Building Styles

Where development is permitted in vulnerable areas and adequate defences cannot be provided, the levels of risk to property owners can be reduced by incorporating specific flood-proofing or ground movement tolerating measures into the building design. In flood prone areas the most effective building modifications include:

(i)   **minimum floor heights**; property can be elevated above a prescribed design flood level either by structural means (stilts) or by raising the property on an earth bund (Figure 11.6). In the Royal Borough of Windsor and Maidenhead, for example, new residential properties on the Thames and Colne floodplains, built on land flooded in 1974, must have an internal ground floor level 0.15m above the 1947 flood level (the highest flood this century).

      Similar policies can apply to areas at risk from coastal flooding. Southampton City Council, for example, have specified that new houses built on low-lying land require floor slabs at 3.4m AoD with all car parks and highways at 3.1m AoD.

(ii)   **means of escape**; Residents of single storey properties are particularly vulnerable to flood events as they cannot escape to safety upstairs as the floodwaters rise. The risks to such individuals can be reduced by requiring a means of escape such as a

Figure 11.6 The location and design of flood-proofed buildings on a floodplain (after Rapanos, 1981).

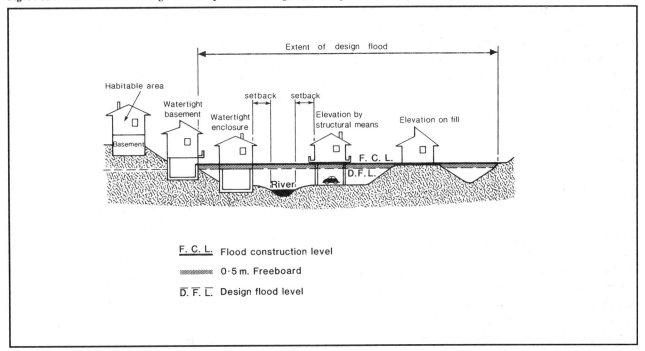

"dormer window" to be incorporated into the building design. Swale BC, for example, have specified that any new houses on land less than 5.3m AoD should contain a means of escape at first floor level, unless the site is protected by secondary defences. In the Romney Marsh area, Shepway DC also require single storey houses, in areas liable to shallow water flooding, to have a means of escape, with no single storey housing allowed in deep water flood areas.

Such measures can be required as conditions attached to planning permissions or detailed in specific development plan policies. However, the Building Regulations do not address flooding problems. Landsliding is considered in the Building Regulations (see Rendel Geotechnics, 1995) and, hence,the building control approval procedures can be used to complement the planning system in areas of coastal instability. This is so in the Isle of Wight Undercliff, where a code of practice for builders is currently being developed which identifies a variety of potentially suitable building styles and techniques that can be successfully employed in areas affected by ground movement (McInnes, 1994; Table 11.9).

Summary

Early warning systems can help reduce the risks facing communities in vulnerable areas by enabling precautions to be taken and evacuation procedures to be initiated before an event occurs. However, they cannot be the sole form of defence for many areas; the onset of an event can be very rapid with little forewarning and for many urban areas the size of the population and value of the property at risk make the potential losses unacceptable. In such circumstances some form of **defence scheme** is needed to reduce the level of risk to an acceptable level. Schemes usually involve a combination of approaches from flood regulation through reservoirs, channel enlargement to the construction of flood walls and embankments. For areas of coastal erosion problems, measures usually involve the protection of the base of the affected cliff and some form of slope treatment measures.

Flood and coastal defences can only provide protection to a certain standard; they cannot eliminate the risk. The standard of protection provided is usually determined by the value of the property at risk; urban areas would normally have a higher standard of defence than rural areas. This is largely a reflection of the costs of such schemes – around £100M is spent each year from the public purse on defence works. Their use should be limited to protecting developed areas at risk, not to create new areas for potential development.

Table 11.9 Suggested scope for code of good practice for building in Ventnor, Isle of Wight (after Lee and Moore, 1991).

---

(1) Siting:

– recommendations intended to avoid unsuitable siting of buildings within a development plot, such as adjacent to a landslide scarp, adjacent to the crest of a steep slope, close to a near vertical face from which rockfalls may occur;

(2) Earthworks:

– the importance of earthworks control in connection with general site preparation and also with landscaping;
– the avoidance of fill operations near the crest of existing slopes, and of excavation at the toe of steep slopes;
– the need for balanced earthworks over the development site;
– restrictions on the length of trenches excavated along the contours of steep slopes;

(3) Retaining walls:

– the avoidance of loading behind, or unloading in front of existing retaining walls, unless the design, construction and condition have been properly investigated and any necessary remedial or strengthening measures carried out;
– advice on the correct design of new retaining walls;
– recommendations covering the adequate consideration of ground water during design, in the detailing of drainage measures, and during construction;

(4) Groundwater control:

– provision for free drainage of groundwater;
– re-routing, repair and reconnection of existing sewers, and water supply network;

(5) Drainage:

– provision for positive drainage of surface water;
– prohibition of septic tanks and soakaways;

(6) Service connections:

– provision of flexible service connections from buildings;
– provision of flexible jointed pipes capable of sustaining small movements without leakage;

(7) Foundation design:

– the requirement for raft foundations, where appropriate, to be designed for potential partial loss of support;

(8) Building form:

– identification of building forms that are unsuitable for landslide areas and advice on those forms that are more appropriate;
– restrictions on height and foundation loading;

(9) Structural form:

– advice concerning both unsuitable and structural forms and those that are more appropriate.

There is, therefore, a clear need for coordination between land use planning and the planning of flood and coastal defence works. In the past, planners have tended to view erosion and flooding issues as technical matters and not specific planning concerns. This has led to a preference and, unfortunately, a necessity to manage flood and erosion risk areas through engineering options rather than planning control. Paradoxically, it may have been the clear success of many defence schemes that has led to a lack of control of development in those areas at risk. For example, construction of defences often leads to increased

pressure for development in what is now perceived to be a safe area. However, increased investment and density of housing behind the defences may only lead to higher losses when, inevitably, larger events occur.

Structural solutions to hazard problems can also lead to conflict with land use planning and other interests in both the river and coastal environment, creating pressure for a "softer" approach which works with natural processes rather than against them. These concerns are central to the overall objectives of English Nature's **Campaign for a Living Coast** (English Nature, 1992):

> "English Nature will seek to halt and reverse the loss of coastal habitats and natural features resulting from coastal squeeze and from the disruption of natural sedimentary systems. We shall try to establish a principle that new or replacement sea defence, coast protection or similar works should not exacerbate coastal squeeze or disruption of systems and **should reverse those wherever possible**," (English Nature, 1992).

Providing fixed defences is only one of a rage of options available to coastal defence managers. Alternatives include retreat to a more easily defensible line: such **managed retreat** options could also provide the opportunity for recreating intertidal habitats (see Chapter 9). Experiments into the viability of this option are currently being carried out at Northey Island in the Essex Marshes, by English Nature, the NRA and the National Trust.

Building modifications provide a further opportunity to ensure that risks to property are minimised. However, the decision to allow new buildings in risk areas can lead to conflict with other flood or erosion management objectives. For example, new properties can increase the risk of flooding elsewhere by generating addition runoff and reducing floodplain storage (see Chapter 10). At present, flooding is not addressed by the Building Regulations.

Chapter 11: References

Barton M.E and McInnes R.G. 1988. Experience with a tiltmeter – based early warning system on the Isle of Wight. In C. Bonnard (ed) Landslides, 379–382, Proc. 5th Int. Symposium on Landslides, Lausanne.

Brampton A. 1992. Beaches – the natural way to coastal defence. In M.G. Barrett (ed) Coastal zone planning and management, 221–220. Thomas Telford.

Brandon T.W. (ed) 1988. Water practice manuals. River Engineering Part II. Structures and Coastal Defence Works. IWEM, London.

Clark A. R. and Guest S. 1991. The Whitby cliff stabilisation and coast protection scheme. In R.J. Chandler (ed) Slope stability engineering: developments and applications. Thomas Telford, 263–270.

Department of the Environment/MAFF/Welsh Office, 1989. Code of Practice on Conservation, Access and Recreation. HMSO.

English Nature 1992. Campaign for a Living Coast. English Nature, Peterborough.

Green C.H., Parker D.J. and Emery D.J. 1983. The real costs of flooding to households: the intangible costs. Middlesex University.

Hershfield D.M. and Kohler M.A. 1960. An empirical appraisal of the Gumbel extreme–value procedure. Journal of Geophysical Research, 65 1737–1746.

Holmes N. 1993. Opportunities and practice: inland works. Proc. MAFF Conference of River and Coastal Engineers.

Hutchinson B.A. 1992. Bradford Beck Flood Alleviation Scheme: Background. Proc. MAFF Conference of River and Coastal Engineers.

HR Wallingford, 1991. A guide to the selection of appropriate coast protection works for geological SSSIs. Nature Conservancy Council.

Leafe R. 1991. The English Nature View. Proc. of the SCOPAC Conference on Coastal Instability and Development Planning, Southsea.

MAFF/DoE/Welsh Office 1991. Conservation guidelines for drainage authorities.

MAFF/English Nature/NRA 1992. Environmental procedures for inland flood defence works.

MAFF/Welsh Office 1993a. A strategy for flood and coastal defence in England and Wales. MAFF Publications.

MAFF/Welsh Office 1993b. Project Appraisal Guidance Notes. MAFF Publications.

MAFF 1993c. Coastal defence and the environment. MAFF Publications.

McIlroy A.J. 1990. Clifftop families admit defeat in battle with the advancing sea. Daily Telegraph 08.10.1990.

McInnes R. 1994. Management of the Ventnor Undercliff landslide complex, Isle of Wight, UK. 7th Conference Ins. Ass Engineering Geologists, Lisbon

NRA 1992. Corporate Plan 1992/93.

Rapanos D. 1981. Floodproofing new residential buildings in British Colombia. Ministry of Environment, Province of British Colombia.

Rendel Geotechnics 1995a. Erosion, Deposition and Flooding in Great Britain. Methodology Report. Open File Report held at the DoE.

Rendel Geotechnics, 1995b. The Undercliff of the Isle of Wight: a review of ground behaviour. South Wight Borough Council.

Smith K. 1992. Environmental hazards: assessing risk and reducing disaster. Routledge.

Tyhurst M.F. 1991. Planning aspects of the Chewton Bunny planning inquiry. Proc. of the SCOPAC Conference on Coastal Instability and Development Planning, Southsea.

Ward R.C. 1978. Floods: a geographical perspective. MacMillan.

West G.M. 1992. Engineering the beaches. In M.G. Barrett (ed) Coastal Zone Planning and Management, 231–236, Thomas Telford.

12 Environmental Management: Erosion, Deposition and Flooding as a Conservation Issue

Introduction

The Government's environmental strategy recognises that the countryside and the coast are a central part of the national heritage (DoE, 1990a). The processes of erosion, deposition and flooding are an integral part of the natural landscapes of those areas and are necessary for creating and maintaining many elements of the landscape that are valuable for recreation, tourism and education, as described in Chapter 2. In this way the processes are important in facilitating the achievement of a number of the Government's environmental policies, most notably:

- to conserve and improve the landscape and encourage opportunities for recreation;

- to give extra protection to areas of special value;

- to conserve the diversity of Britain's wildlife, particularly by protecting habitats.

Sites or areas of national and international conservation value are protected by a wide variety of statutory and non-statutory designations. Conservation objectives can be achieved through a variety of mechanisms, including **management agreements** with owners and occupiers or the **restriction of operations** that could lead to degradation of the areas. In many instances the reliance of a site on regular inundation or continued erosion and deposition can mean that flood and coastal defence works may result in a loss of conservation value. At the centre of this source of potential conflict are a number of key management issues that will be discussed in the following sections: river engineering works, water levels in wetlands, and sediment transport around the coast.

River Corridor Management

River corridors support a wide variety of habitats, many of which are of national importance and, hence, have been designated as SSSIs. Watercourses can also contain important geological and geomorphological features, and are being increasingly used for recreational pastimes such as boating and fishing. However, the flood defences that are needed to protect communities in vulnerable areas can have a major impact on the natural and landscape value of these features.

It is appreciated that, in the past, opportunities for maintaining or enhancing conservation interests were not recognised by many flood defence schemes and, consequently, there has been a significant loss of wildlife habitats (MAFF et al, 1992). Following the significant changes in public and political awareness of environmental issues, the Government and its conservation agencies have emphasised that both flood defence schemes and maintenance works should take account of the need to enhance the natural environment.

Guidelines have been prepared to advise the NRA and IDB's on fulfilling their environmental duties imposed under the Water Resources Act 1991 and the Land Drainage Act 1991 (MAFF/DoE/Welsh Office, 1991). These authorities should consult the relevant interests (see Table 12.1) well in advance of any proposals for new works or improvements to existing works. Consultations should aim to find ways of avoiding damage as well as identifying measures for conserving and enhancing the natural environment and creating new habitats, including riparian buffer zones. **Environmental procedures** for drainage authorities have been prepared jointly by MAFF, the NRA, English Nature and the RSPB which provide a step–by–step guide on how environmental considerations can be addressed at every stage of the decision making process for

Table 12.1 Consultation by drainage authorities for flood defence works (from MAFF/DoE/Welsh Office, 1991).

The NRA and IDBs are strongly advised to consult English Nature, CCW, the Countryside Commission and any other relevant organisation such as English Heritage/Cadw before carrying out any work. They are advised to take particular care where operations may affect directly or indirectly any of the following:

- UNESCO Biosphere Reserves;
- Ramsar Sites;
- Special Protection Areas (SPAs);
- Species protected under the Wildlife and Countryside Act 1981 (as amended);
- Sites of Special Scientific Interest;
- Environmentally Sensitive Areas (ESAs);
- National and Local Nature Reserves;
- Ancient Monuments;
- Listed Buildings;
- National Parks;
- Areas of Outstanding Natural Beauty;
- Heritage Coasts;
- Public Rights of Way; and
- Any other sites of environmental or archaeological interest such as those identified by local authorities and those owned or managed by the National Trust or voluntary conservation agencies.

Other bodies should likewise be consulted where applicable, for example:

- local authorities on conservation, archaeological, landscape and access issues;
- Royal Society for Nature Conservation, Wildlife Trust Partnership;
- Country Sites and Monuments Records Offices;
- National Trust;
- Regional Welsh Archaeological Trusts;
- Council for the Protection of Rural England;
- Council for the Protection of Rural Wales;
- Royal Society for the Protection of Birds;
- Council for British Archaeology;
- Water Services Association Working Party on Industrial Archaeology; and
- Other local amenity and conservation organisations.

carrying out inland flood defence works (MAFF et al, 1992).

Maintenance works can result in environmental damage and, therefore, it is stressed that drainage authorities should consult the conservation agencies well in advance on their intended maintenance programme for the following year (MAFF/DoE/Welsh Office, 1991). This process should seek to establish agreement on the nature, frequency and timing of maintenance operations. However, as the case study of the Dee meanders has demonstrated, conflict may still arise where the proposed designation of a site as a SSSI would require changes to the maintenance programme that could adversely affect flood defence interests (Jones and Campbell, 1993; see Rendel Geotechnics, 1995).

To further its environmental duties (see Table 3.3) the NRA has established a system of **river corridor** conservation surveys as a prelude to flood defence works on main rivers, with the corridor defined as:

"a stretch of river, its bank and the land close by. The width of the corridor depends on how much the nearby land is affected by the river and vice versa. Usually the river corridor includes land and vegetation within 50m of the river bank, but where there are extensive water meadows, marshes, or other wetland areas, the corridor may be wider to include these associated features. The width of the river corridor and the component zones will vary according to channel and floodplain morphology." (NRA, 1992).

Figure 12.1 provides an example of the format of river corridor surveys, as defined by the NRA's handbook on survey practice. Although such surveys are undertaken to allow the NRA to pursue its environmental duties without prejudicial actions from its flood defence activities, the surveys provide a baseline statement against which to monitor environmental changes and may identify opportunities for enhancing the conservation interests of a watercourse (Ash and Woodcock,

Figure 12.1 River corridor surveys: definition diagram and survey example from the River Test (from NRA, 1992).

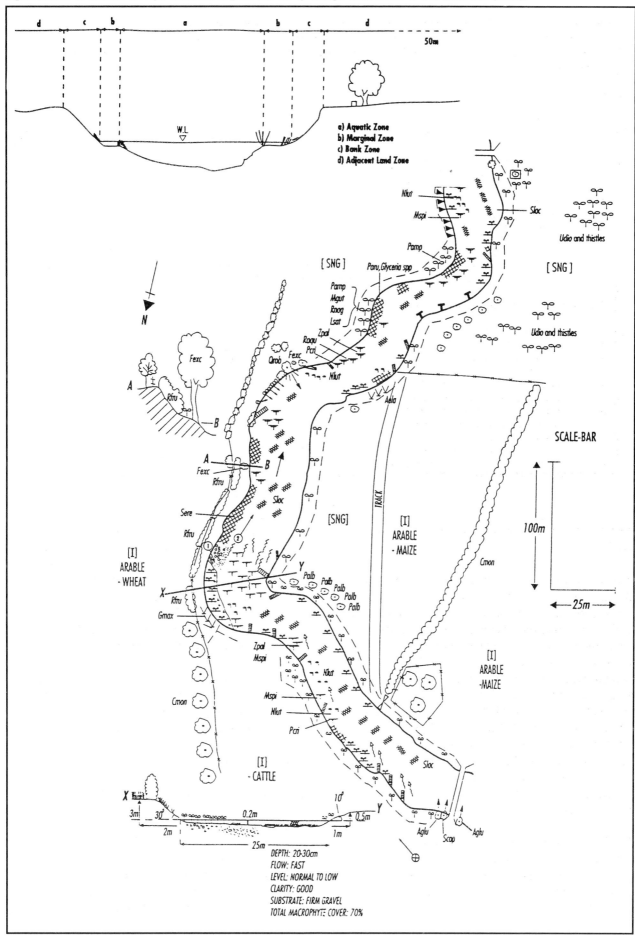

1988). They also provide background information for the development of **Catchment Management Plans** which consider the relationship between flood defence and conservation at a strategic level (see Chapters 3 and 11).

Water Level Management

Wetland habitats can support a wide variety of plant and animal species. Many have been designated as SSSIs, some are of international importance and designated as Special Areas of Conservation, Special Protection Areas, Ramsar sites or World Heritage Sites. These sites are generally dependent on an appropriate water management regime to maintain the different degrees of wetness which are required or tolerated by different species. For example, annual winter flooding benefits over-wintering birds such as wildfowl and waders, but once flood waters subside it is necessary to maintain damp conditions into spring and early summer if breeding waders are to be encouraged. Bankside plant and animal communities are also highly sensitive to fluctuations in water levels. Wetland areas are also important for the wealth of well preserved archaeological remains which they may contain.

Over the centuries, many wetlands have been drained and protected from flooding to create conditions suitable for agriculture and development. Indeed, much of the nation's most productive land is located in these areas. Potential conflict between conservation and flood defence interests has required the preparation of **Water Level Management Plans** by operating authorities (NRA, IDBs and local authorities) in agreement with the conservation agencies, (MAFF/DoE/Welsh Office, 1991; MAFF et al 1994). It is envisaged that the purpose of these plans is to identify conservation opportunities and help ensure that the operating authorities fulfil their environmental duties. Plans are intended to address:

- the aims of the various interests;

- the water level management objectives for the area;

- target water regimes and the range of acceptable tolerance for each water management structure and specify any seasonal requirements;

- contingency measures for exceptional circumstances;

- the timing and nature of maintenance activities which may affect water levels;

- an implementation programme, its phasing and associated cost;

- a programme of monitoring to ensure that the objectives are being met;

- funding and operating responsibilities;

- a review procedure.

Plan preparation should involve extensive consultation with the various interests, from conservation agencies to landowners and recreation groups. MAFF recommend that each consultee should evaluate their own interests and set these in the context of their overall aims for water level management (MAFF et al, 1994). The operating authority should then determine a water level management regime which attempts to reconcile the aims of the different groups. Where management changes are planned it may be necessary to evaluate the associated changes in flood risk. If conflicts remain after consultation the operating authority should prepare a submission setting out **substantive areas of disagreement** which can be referred to the Ministry for advice on how matters might be resolved.

Coastal Management

Disruption of the natural movement of sediment around the coast can have significant consequences for many coastal landforms, as has been emphasised in previous Chapters. Of particular importance is the supply of sediment to sand dunes, shingle beaches, spits and bars, mudflats and saltmarshes from eroding soft rock cliffs. Where coast protection works have prevented or intercepted this supply there have been a number of consequences:

- protected cliffs can lose their importance and interest as geological sites and as elements of attractive coastal landscapes;

- there may be a substantial reduction in the area of a range of coastal landforms from sand dunes to saltmarshes and the loss of much of their geological, wildlife and landscape interest;

- there may be a reduction in the capacity of some coastal landforms to provide effective flood defences. Low lying coasts may, therefore, become dependent on the construction of higher artificial defences to achieve the same standard of protection. This, in turn, can lead to further habitat loss and degradation of geological and landscape value.

Conservation agencies have recognised that the management of coastal assets needs to be based on an understanding of the operation of natural processes operating within large coastal systems (e.g. English Nature, 1992). English Nature, for example, have adopted a presumption in favour of the **unrestrained operation of natural processes**, except where:

(i)   there is an overwhelming case for safeguarding a habitat or feature that is irreplaceable;

(ii)   the local natural systems have been so disrupted or the site so constrained that there is little prospect of re-establishing a non-managed environment.

In an attempt to reconcile potential conflicts with flood and coastal defence interests, conservation agencies are taking an active role in the preparation of **Shoreline Management Plans** for littoral cells or sub-cells (MAFF 1994; Table 4.4). Indeed, the policy of unrestrained operation of natural processes is broadly equivalent to the "do nothing" option that may be the preferred approach to coastal defence on many lengths of unprotected or rural coastlines. Where defences are considered to be necessary, it is now clear that they should be environmentally acceptable; in general, schemes will not be approved by MAFF or the Welsh Office if they are considered unacceptable by the conservation agencies. Indeed, the potential impact of a scheme on habitats and the environment are key considerations in the appraisal procedures for defence works (MAFF, 1993a, 1993b) as highlighted by a **presumption against the disruption of natural processes** except where life or important man-made or natural assets are at risk (MAFF, 1993a). It is hoped that these recent developments will provide an effective mechanism for reconciling coastal defence interests with the needs of nature conservation and help secure English Nature's objectives in seeking to half and

reverse the loss of habitats and coastal features, and to maintain these features in a sustainable condition (English Nature, 1992).

The Role of the Planning System

The planning system is an important instrument in achieving conservation objectives above LWM, through the inclusion of land use policies within development plans that take into account the following guidance:

- PPG 7 The Countryside and Rural Economy;

- PPG 9 Nature Conservation;

- PPG 20 Coastal Planning.

Concern has been expressed, however, over the inconsistent or inadequate treatment of many nature conservation objectives within development plans. In 1990, for example, RSPB published a review of nature conservation policies included within structure plans, for England and Wales (Bain et al, 1990). This document also included an analysis of threats posed to Ramsar and SPA sites by proposals within the plans. As a result of their analysis RSPB identified policies aimed at protecting sites of known conservation interest in all plans. However, RSPB found that there was a wide variation of wording used in these policies, with 22 out of 56 structure plans having what they described as "inadequate" SSSI policies. This variation was felt to weaken the conservation case when objecting to planning applications:

> "When a structure plan is reviewed, there is a risk that developers will try to exploit weak site safeguard policies in neighbouring structure plans by recommending their adoption by the county council as 'reasonable and pragmatic policies'; councils must guard against this practice for site safeguard. Poorly worded and therefore ineffectual site safeguard policies do not provide a firm enough foundation on which conservation bodies can built a defence for a conservation site", (Bain et al, 1990).

Despite the observed inconsistencies, the planning system remains an important source of protection for conservation interests. For example, the system has proved to be effective in arresting the spread of development in the countryside and along the undeveloped coast. This role can be further strengthened by ensuring that planning decisions are consistent with site management objectives and the inclusion of policies in development plans that support the aims of the relevant Catchment, Water Level or Shoreline Management Plans.

Summary

Reconciling flood and coastal defence interests with the need to enhance and improve the natural environment is a potential source of conflict as the very processes which pose a threat to vulnerable communities in one location can be essential for maintaining valued conservation features elsewhere. Management plans, however, provide a mechanism for resolving these conflicts between operating authorities and conservation interests. The plans are non-statutory and based on extensive consultation; it is intended that they should present an agreed management strategy that is compatible with maintaining conservation features in a sustainable condition.

Both the NRA and IDBs have environmental duties which oblige them to further conservation interests in undertaking flood and coastal defence works. In this context the Government, in association with conservation agencies and groups, has provided guidance to these operating authorities on the environmental procedures that should be carried out when undertaking works (MAFF/DoE/Welsh Office, 1991; MAFF et al, 1992; MAFF, 1993c). Once again the emphasis is on extensive consultation with statutory and non-statutory organisations in order to reach agreement at an early stage of the planning defence schemes and maintenance programmes. All grant-aided schemes in England and Wales must be environmentally acceptable (MAFF, 1993a).

The planning system can have an important role in supporting conservation objectives, most notably through the restriction of development that could affect conservation features and the inclusion of policies within development plans that support the aims of the various Catchment, Water Level and Shoreline Management Plans. Since 1947, the system has proved to be effective in arresting the spread of piecemeal development in the countryside and along the undeveloped coast.

The preceding discussion has concentrated on the recent developments in conservation management

that have taken place in England and Wales. Similar observations apply to Scotland although, with the exception of planning guidance, there have been no comparable initiatives for reconciling conservation interests and the risks from erosion, deposition and flooding.

Chapter 12: References

Ash J.R.V. and Woodcock E.P. 1988. The operational use of corridor surveys in river management. Journal of the Institution of Water and Environmental Management, 2, 423–428.

Bain C., Dodd A. and Pritchard D., 1990. RSPB Planscan. A study of development plans in England and Wales. RSPB Conservation Topic Paper No. 28.

Department of the Environment 1990. This Common Inheritance. Britain's Environmental Strategy. HMSO.

English Nature 1992. Campaign for a Living Coast. English Nature, Peterborough.

Jones R. and Campbell S. 1993. The Dee meanders – case of Dee – stabilisation. Earth Heritage 1, 13–15.

MAFF, DoE and Welsh Office 1991. Conservation guidelines for drainage authorities. MAFF, English Nature and NRA, 1992. Environmental procedures for inland flood defence works. MAFF Publications.

MAFF 1993a. Strategy for flood and coastal defence in England and Wales. MAFF Publications.

MAFF 1993b. Project Appraisal Guidance Notes. MAFF Publications.

MAFF 1993c. Coastal defence and the environment. MAFF Publications.

MAFF 1993d. Shoreline management plans. A guide for operating authorities. MAFF Publications.

MAFF, Welsh Office, Association of Drainage Authorities, English Nature and NRA, 1994. Water level management plans: a procedural guide for operating authorities. MAFF Publications.

NRA 1992. River corridor surveys. Conservation Technical Handbook 1. NRA, Bristol.

Rendel Geotechnics, 1995. Erosion, Deposition and Flooding in Great Britain. Methodology Report. Open File Report held at the DoE.

13 The Planning System as an Instrument for Management of Erosion, Deposition and Flooding Issues

Introduction

From the preceding discussion of the range of available approaches for managing erosion, deposition and flooding–relating issues, it is clear that the **planning system** can be used in a variety of ways to achieve management objectives, including:

- ensuring that certain river channel or foreshore maintenance works do not have an adverse effect on environmental interests (Chapter 8);

- ensuring that new development does not restrict access to watercourses for channel maintenance works (Chapter 8);

- avoiding locating new development in unsuitable areas or specifying restrictions on housing occupancy in risk areas (Chapter 9);

- discontinuing or restricting existing land uses or buildings in vulnerable areas through discontinuance orders or compulsory purchase and providing compensation to affected parties (Chapter 9);

- ensuring that precautions are taken to prevent runoff from new developments increasing flood risk (Chapter 10);

- ensuring that development does not adversely affect floodplain storage and, hence, increase flood risk (Chapter 10);

- ensuring that development does not affect coastal cliff stability or lead to an increase in coastal erosion (Chapter 10);

- ensuring that coast protection works and new flood defence works are compatible with land use planning and conservation objectives in an area (Chapter 11).

- supporting environmental management objectives by ensuring that development does not lead to a decline in value of the conservation sites dependant on the continued operation of physical processes (Chapter 12).

However, there is a difference between what **can be achieved** and what **has been achieved** by local planning authorities. This is a reflection of the prevailing view, until the mid 1980's, that hazards were a matter for the developer and not a planning issue. Although recent planning policy advice in England and Wales (PPG 14; PPG 20 and Circular 30/92) has emphasised the need to treat flooding and coastal erosion as **material considerations**, it is clear that the opportunities provided by planning systems have not been fully exploited. Table 13.1, for example, summarises the results of a survey of a sample of local planning authorities in a selection of different regions of Great Britain where potential problems are known to occur. Different local planning authorities have responded to the presence of significant flooding and coastal erosion problems within their areas. It is apparent that there has been a wide range of planning responses, ranging from the sophisticated strategies for both flooding and erosion risks in Norfolk and Suffolk to no planning policies in many areas of South West England, Wales and Scotland (Rendel Geotechnics, 1995). When considering individual policies a number of observations can be made:

- 50% of the authorities have policies directed towards **avoidance** of vulnerable areas, rising to over 75% in South East England;

Table 13.1 A summary of erosion, deposition and flooding–related development plan policies in a sample of local planning authorities in 5 regions of Great Britain.

	South East England	Yorkshire and Humberside	South West England	Wales	Scotland	TOTAL
Number of Local Authorities responding	48	6	20	18	30	122
Maintenance Works	1	0	0	0	0	1
Avoidance	35	3	4	11	6	59
Occupancy restrictions	1	0	0	0	0	1
Discontinuance	0	0	0	0	0	0
Prevent effects of Runoff	11	1	1	1	0	14
Floodplain Storage	10	1	0	0	0	11
Effect on Cliff Stability	2	0	0	0	0	2
Effect on Sediment Transport	0	0	0	1	0	1
Flood and Coastal Defence Works	0	0	1	0	0	1
Building Modifications	8	1	1	0	0	10
No relevant Policies	12	2	16	7	24	61

Note: The overwhelming majority of development plans examined had policies which specifically addressed conservation issues.

- most other policies are only occasionally included in development plans, although the evidence suggests that restrictions on the generation of surface water runoff and reduction in floodplain storage are becoming widespread in parts of the South East England;

- no examples were identified of planning policies directed towards the discontinuance of land use or compulsory purchase in vulnerable areas.

It is important to recognise, however, that not all areas with flooding or coastal erosion problems have been developed or experience the same pressures for development. Thus, the absence of specific planning policies in areas such as Grampian, Powys, South Glamorgan and North Yorkshire may be seen, in part, to reflect that erosion, deposition and flooding are not priority issues. In many cases, undeveloped areas of river corridors or coast can have strong protection against unsuitable development through statutory conservation designations or non–statutory designations and associated planning policies which aim to divert development away from these areas, i.e. **avoidance**.

Model Standards of Good Practice

The reasons why the planning system does not appear to fully exploit the opportunities for contributing to the management of erosion, deposition and flooding issues are likely to be complex, reflecting a combination of the national context set by the legal and administrative framework, local planning authority perceptions of the issues, the attitudes of planning inspectors and local circumstances. In the following sections an attempt is made to identify some of the key factors which appear to constrain the effectiveness of the planning system. In order to achieve this it is necessary to set out a model of good practice and seek to examine how current arrangements are effective in helping the system meet this model. The seven model standards set out below reflect the central themes of environmental management, namely advance warning of potential problems, the need to take account of the dynamic nature of the physical environment and the importance of liaison between planning and other management systems:

1.   the nature of the processes and associated issues varies across different terrain units, i.e. the system should be able to take account of **geographical variability**;

2.   erosion, deposition and flooding are interrelated processes operating within large physical systems (catchment and coastal systems); the operation of a process in one area may have an impact over a very large area, affecting the level of risk experienced elsewhere within a physical system, i.e. the planning system should be able to take a **strategic approach** to these issues;

3.   erosion, deposition and flooding are natural processes that can have important roles in creating and maintaining national resources such as nature and geological conservation sites, fisheries spawning grounds, aggregate resources, etc., i.e. the system should be able to take a **resource-based perspective**;

4.   development and land use can have significant effects on the operation of physical processes and the degree of risk elsewhere, i.e. the system should be able to take a **precautionary approach** to hazard management;

5.   the nature of the processes and associated risks may vary over time, especially in the river and coastal environments, i.e. the system should be able to take account of **temporal variability**;

6.   the planning response to erosion, deposition and flooding issues should be consistent between neighbouring areas and coordinated with other approaches to catchment and coastal management, i.e. the system should be able to take an **integrated approach** to hazard management;

7.   the planning system should consider the issues associated with erosion, deposition and flooding processes within the broad framework of land use issues in an area, rather than in isolation, i.e. a **balanced approach**.

The Legal Framework

The principal planning legislation (the Town and Country Planning Act 1990; the Town and Country Planning (Scotland) Act 1972) defines a flexible framework within which local planning authorities can achieve most of the model standards outlined above, through the inclusion of policies within development plans and the determination of planning applications. Amongst the most important provisions are:

● local planning authorities shall keep under review matters expected to affect the planning and development of their area (1990 Act S.11, S.30). This shall include the principal physical characteristics of the area and, so far as they may be expected to affect that area, of any neighbouring areas, i.e. allows model standards 1, 2, 3, 5 and 7 to be met;

● structure plan policies should have regard to current policies for **regional** economic planning and development (1990 Act S.31), i.e. model standard 2;

● local plan proposals should be in general conformity with the relevant structure plan (1990 Act S.46), i.e. model standard 6;

● in dealing with planning applications local authorities must have regard to the development plan and **any other material considerations**, i.e. can allow model standards 1–7 to be met.

There are, however, limits to the influence of the planning system:

(i)   the system is designed to regulate **development**, i.e. "the carrying out of buildings, engineering, mining or other operations in, on, over or under land, or the making of any material change in the use of any building on other land" (The Town and Country Planning (Scotland) Act 1972 S.19; The Town and Country Planning Act 1990 S.55). Under the 1990 Act a number of activities are defined as not to constitute "development" of land and, therefore do not require planning permission (S.55). These include:

–   maintenance, improvement or alteration of a building which does not materially affect its external appearance;

–   the carrying out of road improvements or maintenance works on land within the boundary of the road;

147

- the carrying out of inspection, repair or renewal works (e.g. for sewers or water mains) by local authorities or statutory undertakers;

- the use of a building or land within its curtilage for any purpose incidental to the enjoyment of the dwelling;

- the use of land and existing building for agriculture and forestry.

(ii) the administrative area of most local authorities and, hence, the limit of planning control, is normally around low water mark, although there are notable exceptions such as the Shetland Isles whose administrative area extends to the limit to territorial waters (under the Zetland County Council Act 1974);

(iii) not all development requires specific planning permission. In England and Wales, the **General Development Order 1988** (GDO; as amended) gives general planning permission in advance for certain defined classes or development set out in Schedule 2 to the Order, i.e. **permitted development**;

(iv) most **harbour authorities** have a significant degree of autonomy to operate under statutory powers set out in local Acts of Parliament or by means of orders made under the Harbours Act 1964.

It is clear that the planning system does not have control or a direct influence over all land or coastal zone uses that may affect the occurrence and significance of erosion, deposition and flooding processes. Amongst the more important activities beyond planning control are:

- agriculture and forestry operations;
- changes in land management practices, e.g. forest clearance, land drainage;
- construction of forest roads and tracks;
- construction of swimming pools, terracing of gardens, open trench excavations, removal of vegetation from slopes and building improvements;
- improvements, maintenance or repair of watercourses or land drainage works;
- development on harbour authority or NRA operational land;

- taking up, diverting or altering the level of harbour watercourses;
- dredging operations and spreading of dredged spoil on operational land;
- building improvements and construction;
- marine aggregate extraction;
- development below LWM.

It is important, therefore, for planners to coordinate their activities with authorities or bodies with responsibilities for other regulatory systems (e.g. NRA land drainage consents, the Government View Procedure for marine aggregate extraction, etc; see Chapter 3).

Planning Guidance

Although legislation sets the framework for the planning system, the way it operates is strongly influenced by **policy advice** from the Government. National Guidance on relevant planning considerations is issued in a variety of ways, including Regional Planning Guidance, Circulars, Planning Policy Guidance and National Planning Guidelines (see Chapter 3). These documents are important in helping local authorities define "material considerations" and can be viewed as benchmarks against which development plans and decisions on planning applications can be judged. In this sense, current planning guidance presents the Government's view of the planning system's role in managing erosion, deposition and flooding issues and sets a standard that local planning authorities are expected to meet.

Current planning advice regarding certain erosion, deposition and flooding issues has been incorporated in PPG 14, PPG 20 and Circular 30/92 (DoE, 1990; DoE, 1992a and DoE 1992b, respectively) which apply in England and Wales; no specific advice on these matters has been issued in Scotland. The guidance has been issued since 1990 and can, therefore, be regarded as an up-to-date Government view. Whilst it addresses various management aspects of flooding and coastal hazards (Table 13.2) it does not, however, cover many of the model standards, as will be described below:

(i) **geographical variability**; as was emphasised in Chapter 4, lack of awareness of the potential for damaging events can be a significant problem for local planning authorities, especially where the threat is associated with extremely rare events or

Table 13.2 A summary of management strategies addressed in current planning guidance in England and Wales.

MANAGEMENT STRATEGY	ADDRESSED BY PLANNING GUIDANCE	COMMENT
Operation of maintenance operations	No	
Ensure access for maintenance operations	Yes	**Circular 30/92** advises local authorities to use their powers to restrict development that would interfere with the NRA's ability to carry out maintenance (para 4) and draws attention to the way NRA may respond to development control consultation by suggesting conditions to ensure access for maintenance (para 16).
Avoidance	Yes	**PPG14;** coastal authorities may wish to consider presumption against built development in areas of coastal landslides or rapid coastal erosion (para 29). **Circular 30/92;** advises local authorities to guide development away from areas that may be affected by flooding (para 4). The NRA may identify set back lines on some coasts beyond which most development should be avoided (para 7). **PPG20;** policies in vulnerable areas should be to avoid putting further development at risk (para 2.13). Development should not be allowed to take place on coastal cliffs where erosion is likely to occur during the lifetime of the building (para 2.16).
Occupancy Restrictions	No	
Discontinuance and Compulsory Purchase	No	
Runoff Control	Yes	**Circular 30/92;** notes that new development, some mining, land drainage and forestry development can increase the quantity and rate of runoff reaching watercourses, causing an increase in flood risk (para 17). The guidance advises on the need for suitable alleviation works prior to development of the site (paras 18–20).
Floodplain Management	Yes	**Circular 30/92;** notes that local authorities should use their powers to restrict development that would increase the risk of flooding (para 4). Consultation arrangements with the NRA may need to make special provision for developments which could result in significant loss of floodplain storage (para 13).
Cliff Management	No	**PPG14;** does note however, that the main cause of contemporary landslide movement is human activity (para A49).
Sediment Transport Considerations	Yes	**PPG20;** notes that the impact of coastal defences and other developments on the movement of coastal sediment should be taken into account (paras 2.18 and 4.3).
Conservation and Defence Works	Yes	**PPG14;** notes that stabilisation works may, by their size and location, invoke the need for environmental assessment (para 29). **Circular 30/92;** notes that the close relationship between flood defence and nature conservation may demand consultation with conservation agencies (para 11) and draws attention to "soft" alleviation measures for runoff control (para 20). **PPG20;** notes the effect of coastal defences on areas of conservation value (para 4.3).
Building Modifications	No	

failure of man-made structures. The first step of any **general assessment** to provide background information in support of planning decisions is to recognise the types of problems that can be anticipated in particular terrains and them to identify the settings where problems could occur. The current guidance is not directed towards highlighting how the nature of erosion, deposition and flooding problems vary between different settings, with the exception of Circular 30/92 which notes the important distinctions between inland and coastal flooding. Many of the vulnerable settings identified on the 1:625,000 scale thematic maps accompanying the Report are not drawn to local planning authorities attention by the existing guidance, as highlighted in Figure 13.1.

(ii)   **strategic approach**; whilst current guidance is directed towards the most significant problems (i.e. lowland flooding, coastal cliff erosion and coastal flooding) it does not address many processes that can have important indirect impacts or lead to an increase in risks elsewhere. Figure 13.2, for example, indicates the extent to which the guidance highlights the various impacts associated with the different processes; many of these impacts can raise significant land use planning issues such as the effect of erosion and sedimentation on flood risk.

PPG 20 Coastal Planning draws attention to the need to consider the impact of coastal defences on the "natural movement of material around the coast" (DoE, 1992a, para 2.18), and the broad scale of operation of coastal processes (paras. 4.2 and 4.11). However, Circular 30/92 does not highlight the importance of considering inland flooding within the context of catchments or as a strategic issue that needs to be considered at a regional level. In a similar way, PPG 14 tends to present coastal landsliding largely as a local issue, rather than emphasising the contribution the process can make to regional sediment budgets. There is also an absence of advice on the need to consider the operation of the processes as a strategic issue in current Regional Planning Guidance.

(iii)   **resource-based perspective**; although PPG 14 notes that instability may be associated with sites of nature and geological conservation value, it does not indicate the extent to which such sites may be dependant on the continued operation of coastal erosion (DoE, 1990; para. 44). In both Circular 30/92 and PPG 20 natural processes tend to be presented as risks to land use and development, rather than drawing attention to their positive role in creating and maintaining river corridor and coastal environments. Where the sensitive nature of these environments is described, it is in the context of the effects of flood and coastal defence works and not as a positive feature adding to the character of an area.

(iv)   **precautionary approach**; the effects of development on local amenities, conservation interests and the environment have always been a factor in determining planning applications. Many of the potential effects on the operation of physical processes may occur well away from the development site or take many years to become apparent, and, hence, present difficult problems to planning authorities. In general, the current guidance provides a clear indication of many of the potential effects of development (Figure 13.3) with notable omissions including:

● the need for improvements in dam safety if development is allowed in some rural valleys;

● the effects of flood defences on flood risk elsewhere, caused by the loss in flood wave attenuation;

● the effects of development on coastal cliff instability problems, especially the significance of water leakage, uncontrolled surface water discharge and inappropriate excavations.

(v)   **temporal variability**; Circular 30/92 notes that in coastal areas marine erosion can rapidly remove areas of high ground protecting lower lying areas behind and, hence, coastal lands should **always** be regarded as being at some degree of risk of flooding (DoE, 1992b; para. 7). In both PPG 14 and PPG 20 the potential for changing levels of risk is implicit in the general comments about the effects of development on neighbouring areas (DoE, 1990; para. 22; DOE, 1992a; para. 213), although neither draw attention to how the level of risk may alter as a result of the cumulative effects of natural processes.

Although both Circular 30/92 and PPG 20 draw attention to the need to take account of rising sea levels and global warming, neither indicate how these changes could modify the nature of hazard in an area or increase the area at risk. PPG 14 does not consider the effects of climate change on the occurrence of instability problems.

(vi)   **integrated approach**; although the central theme of Circular 30/92 is the importance

Figure 13.1 The extent to which the current planning guidance in England and Wales identifies vulnerable settings for erosion, deposition and flooding problems.

Process	Vulnerable setting	PPG 14	Circular 30/92	PPG 20
Upland soil erosion	● peaty soils			
	● areas of recreation pressure			
	● recently afforested land			
Water erosion	● silty and fine sandy soils			
	● areas of poor land management			
Wind erosion	● fine sandy soils and lowland peats			
	● areas of poor land management			
Mudfloods	● dry valleys			
Flash floods	● adjacent to upland streams		■	
	● adjacent to lowland tributaries		■	
	● inadequate storm drains			
	● downstream of dams			
Unstable river channels	● margins of upland areas			
	● channelisation works, etc		■	
Lowland floods	● floodplains			
Sedimentation	● lakes and reservoirs			
	● alluvial rivers			
	● esturies			
Coastal floods	● low lying land		■	■
	● estuaries		■	■
Coastal cliff erosion	● unprotected soft rock cliffs	■		
Wind erosion	● unstabilised sand dunes			

of liaison between local planning authorities and the NRA, the guidance does not draw attention to the role of catchment management plans or the need for coordination between authorities within the same catchment. PPG 20, in contrast, recognises the importance of liaison with coastal defence groups (para. 4.2) the need for regional planning conferences to take a strategic view of coastal defence issues (para. 4.11) and for neighbouring counties to consult during the preparation of their development plans (paras. 4.1 and 4.13). In this document the Government encourages **cooperative working** between local authorities around estuaries and on stretches of the open coast, emphasising the need to involve fully other agencies and bodies with an interest in the coast (para. 4.4). PPG 20 also outlines the relationship between estuary or coastal management plans and development plans (para. 4.17).

PPG 14 does not address the need for an integrated approach to coastal cliff management or draw attention to the role of coast protection authorities and coastal defence groups (some were established prior to 1990).

(vii) **balanced approach**; all the relevant planning guidance is directed towards encouraging local planning authorities to define vulnerable areas and set out policies to ensure that such areas are avoided by future development or that appropriate

Figure 13.2 The extent to which current planning guidance in England and Wales highlights the impact of erosion, deposition and flooding.

PROCESS	IMPACT	PPG 14	Circular 30/92	PPG 20
Upland soil erosion	● land degradation, loss of amenity			
	● loss of water storage capacity in reservoirs			
	● decline in water quality			
	● sedimentation and increased flood risk			
Water erosion	● loss of productivity			
	● long-term decline in yields			
	● sedimentation and increased flood risk			
	● mudflood events			
	● decline in water quality			
	● inconvenience to road users, etc.			
Wind erosion	● loss of productivity			
	● long-term decline in yields			
	● sedimentation and increased flood risk			
Mudfloods	● damage to property			
	● blocked roads and drains			
	● loss of productivity			
Flash floods	● widespread erosion and deposition			
	● potential for loss of life and injury		■	
	● property damage, inconvenience etc.		■	
	● creation of conservation sites			
Unstable river channels	● loss of land			
	● scour around bridges, embankments etc.			
	● creation and maintenance of conservation sites		■	
Lowland floods	● potential for loss of life and injury		■	
	● property damage, inconvenience etc.		■	
	● maintenance of floodplain habitats			
Sedimentation	● loss of reservoir capacity			
	● increased water supply costs			
	● increased flood risk			
	● navigation problems			
	● need for disposal sites for dredging			
	● creation and maintenance of saltmarshes etc.			
	● maintenance of natural coastal defences			
Coastal floods	● potential for loss of life and injury		■	■
	● property damage, inconvenience etc		■	
	● maintenance of conservation sites		■	
Coastal cliff erosion	● loss of land	■		■
	● potential for loss of life and injury	■		■
	● damage to property, inconvenience etc.	■		■
	● supply of sediments			■
	● creation and maintenance of conservation sites	■		
Wind erosion in dunes	● burial of land			
	● inconvenience and disruption			
	● creation and maintenance of conservation sites			

Figure 13.3 The extent to which current planning guidance in England & Wales identifies potential effects of development on the operation of erosion, deposition & flood processes.

System	Potential Effect of Development	PPG 14	Circular 30/92	PPG 20
Hillslope	● increase in runoff and increased flood risk		■	
	● creation of urban flood problems		■	
	● effects of channelisation		■	
	● bridges and culverts acting as temporary dams		■	
	● effect of mining and forestry on flood risk		■	
River	● required improvements in dam safety			
	● changes in channel size due to changes in flows		■	
	● effects on habitats and geomorphological features		■	
	● reduction in floodplain storage and infiltration		■	
	● loss of floodwave attenuation			
	● dredging may increase sedimentation			
Coast	● disruption to sediment supply and transport			■
	● effects of mineral extraction on level of risk			■
	● inter-tidal squeeze			■
	● uncontrolled surface water discharge on slope stability			
	● leaking water pipes and sewer systems			
	● effects of inappropriate excavation on slope stability			
	● recreational pressure on dune stability			

precautions are taken. In this sense the guidance **directly** addresses the issues and promotes the use of earth science information to support decision-making. For example, PPG 20 states that development should not be allowed in coastal areas where erosion is likely to occur during the lifetime of the building; such areas should be clearly defined in development plans. In response to this guidance, Waveney D.C. have established a predictive line along their coastline representing the estimated position of the coastline in 2068. The line was defined from historical cliff recession rates, taking into account the effects of sea level rise (Waveney D.C., 1993; Tyrrell, 1994). The "set-back" line has generated considerable concern within the local community especially in those areas where there is a policy not to provide coast protection works.

It is important to recognise that cliff recession is an irregular process and the "rate" may depend on the timescale chosen. Indeed, variations in historical recession rates may lead to difficulties in precisely defining set-back lines within specific confidence limits (see Chapter 6). In this context, MAFF have recently commissioned a three-year programme of research into developing methods for predicting soft cliff recession rates. Preliminary results from this study suggest that sufficiently accurate predictions should involve a combination of historical assessment, geomorphological appraisal and empirical or probabilistic modelling of the cliff, shoreline and marine processes. In this context the degree of precision required to confidently establish set–back lines may be obtainable only for relatively short distances or eroding cliffs rather than whole stretches of coast.

Policies which need to be supported by accurate predictions of soft cliff recession rates are probably best suited to stretches of the developed coast or where development is being considered. Elsewhere development can be controlled through the use of landscape or nature conservation policies, as was noted in Chapter 12. In addition to being a more effective use of resources, the different approach in developed and undeveloped areas has the advantage of taking a balanced view of natural processes; they are an integral part of the physical environment and only become hazards where development encroaches into vulnerable areas. Current guidance does not, however, promote this view tending to present the processes as problems, both on the coast and in river environments.

From the preceding discussing it is clear that current planning guidance in England and Wales does not fully set out a broad framework of advice that indicates how local planning authorities can utilise the existing planning powers to minimise problems associated with erosion, deposition and flooding processes. It should be emphasised that the guidance does address the most significant problems (flooding and coastal erosion) and establishes the important principle of avoiding areas at risk. There are, however, a number of significant deficiencies within the existing guidance, most notably:

- the nature and extent of hillslope erosion, deposition and flooding processes, some of which can lead to increased risks in the river environment;

- the significance of river channel deposition and land use planning issues raised by dredging operations;

- the role of discontinuance orders and compulsory purchase as a planning response to flood and coastal erosion risks;

- the use of building modifications and occupancy restrictions to minimise risks;

- the strategic nature of erosion, deposition and flooding processes within river catchments;

- the importance of erosion, deposition and flooding in creating and maintaining natural habitats and geological features;

- the role of conservation policies in managing erosion, deposition and flooding issues in undeveloped areas;

- the effects of development on coastal cliff erosion;

- the need for a coordinated approach to catchment management between local planning authorities and the NRA;

- the relationship between catchment management plans and the development plan process.

The absence of specific guidance in Scotland may be a significant constraint on the use of the planning system to minimise the risks from erosion, deposition and flooding. Although problems are generally less severe than in parts of England and Wales, they do occur; the 1993 floods in Tayside resulted in an estimated £12M of damage in the City of Perth. There is a need to develop a clearer planning response to such problems to ensure a consistent and compatible approach between different local planning authorities. Consideration should be given to the need for planning guidance on this issue (NB: Draft guidance for consultation was issued by the Scottish Office in March 1995).

Local Planning Authority Perceptions

The model standards for the planning system, set out above, require local planners to have a level of awareness of the physical processes operating within their area and the issues that can be associated with the management of the physical environment. It has been suggested, however, that many local planning authorities have not given due weight to these factors alongside other considerations such as amenity, conservation and socio–economic factors (Brook and Marker, 1988). Indeed, the principal development constraints in many areas tend to be considerations of landscape amenity and conservation; traditional planning constraints that are well understood by the professional and considered more easily defended than constraints based on hazard. As a consequence, in balancing a wide range of planning objectives, local planning authorities may consider risks from erosion, deposition and flooding to be less important than other benefits such as the conservation of the nature environment. Indeed, conflicts have been identified where the presence of extensive areas of designated conservation value has led to pressure for development on undesignated areas where the risks from physical processes are significantly higher, (e.g. in the Thames Catchment; see the lower Kennett Valley case study in the Methodology Report; Rendel Geotechnics, 1995)

It should be recognised that many planning decisions are strongly linked to social, economic and environmental factors. There are situations where the nature of the physical environment is very relevant to safe, cost–effective development of land. However, in the absence of guidance to the contrary many local planning authorities have, in the past, taken the view that natural hazards such as erosion, deposition and flooding were technical matters. Indeed, the situation prior to mid 1980's can be summarised by:

> "it is generally held that economic considerations e.g. the feasibility of the proposed development, is a matter for the developer, not for the planning authority. In this connection it could be considered that the extra costs which should be incurred in site investigation, land stability and protection works are not land use planning matters", (J.S. Turner, Planning Appeals Commission, 1987).

Another important consideration is reported to have restricted the use of planning controls in vulnerable areas is the right of a landowner to serve a **purchase notice** for land where planning permission is refused and which no longer has a "beneficial use". This could prove expensive for local authorities and may lead to the authority becoming the owner and manager of "useless" pockets of land. The threat of such action is reported to have been enough to sway a decision (Turner, 1987).

The decision to permit development in flood risk areas or on eroding coastlines may, of course, be the result of a planning appeal against a refusal by the local authority. Neal and Parker (1988) provide evidence of a number of major developments on the Thames floodplain at Datchet, that were granted planning permission on appeal. For example, in considering a housing estate within the area affected by the 1947 floods (which could lead to a displacement of floodwater over 10,000m^3 in a similar event), the Planning Inspector found that flood risk was **insufficient reason for refusal**, despite the recognised importance of the flooding considerations:

> "It is accepted that the quality of the appeal site to store water as part of the washlands would be reduced by the proposed development of the site and that this must be taken into account by the Secretary of State. It is accepted that this proposed development is contrary to the advice contained in Circular 52/62, in that there could be a marginal increase in danger to other people by reducing the area of washland; it is also accepted that the aggregate effect of such development could be serious". (DoE, 1979).

In another instance, permission for twenty six houses on the floodplain at Datchet was granted in 1978 on appeal because of the presence of nearby earlier developments were considered to rule out flooding as a factor in reaching a decision (Neal and Parker, 1988). These decisions, and may others around the country, appear to highlight the way in which natural hazards have been viewed by the DoE's Inspectors as less important factors than, for example, the provision of housing or employment opportunities in an area. In such circumstances it should not be surprising that some local planning authorities have taken a similar view.

However, since the mid 1980s there has been a notable change in emphasis given to the problems associated with natural processes. These changes are a reflection of:

(i)   increasing concern about the potential effects of sea level rise;

(ii)   clear statements of national planning policy in England and Wales, (i.e. PPG 14; PPG 20 and Circular 30/92);

(iii)   responses to major hazard events, (e.g. the Towyn Floods of February 1990; the Tayside floods of 1990 and 1993; the Holbeck Hall landslide in June 1993).

Local planning authority responses will, inevitably, lag behind these important events and changes, especially as many authorities would wait until the next revision of their development plan before preparing specific policies. The extent to which these events influence the operation of the planning system will only become apparent over the next five years or so.

It should not be assumed, however, that local planning authorities only follow the course of events and changing Government perceptions. They can have an important role in shaping future policy. Sefton MBC, for example, has recently pioneered many coastal planning policies (Table 13.3) which were influential in shaping the content of PPG 20 Coastal Planning (Sefton MBC, 1989; 1991).

Such awareness of the importance of physical processes is rare amongst local planning authorities. Brook and Marker (1987, 1988) have suggested that this is, in part, a reflection of the inaccessibility of much scientific research and basic earth science mapping to planners, especially those without a background in these disciplines. Indeed, there has been a move away from geology being an element in the formal training of planners; the RTPI, for example, no longer include a compulsory section on geology as part of its professional qualification.

Access to technical earth science information will often be limited to site investigation reports prepared by a developer in support of a planning application. In many instances such proposals will be considered in relation to planning policies developed with little or no regard for the physical processes and, as a result, planners without a background awareness of the significance of these processes may not be aware that the proposal could have considerable effect on property or environmental interests elsewhere within a catchment or coastal system.

It is recognised by the DoE that a mechanism is needed to enable the wide range of physical processes and ground conditions to be taken into account by local planning authorities, (e.g. DoE, 1991). To this end **Applied Earth Science Mapping** techniques that have been developed and promoted by the Department of the Environment over the last decade or so. The techniques involve collecting and collating available earth science information from various sources and then summarising it in a form specifically tailored to meet local planning needs. In general, this type of study is directed towards preparing a combination of thematic maps at a general scale, which become increasing focused on key planning issues as they are developed from the basic factual information (see Chapter 14).

Despite the success of Applied Earth Science Mapping in areas where the DoE have commissioned such studies (e.g. Torbay; Boothroyd, 1991), many of the local planning authorities contacted during this study appear unaware of the techniques. This, in part, is due to an absence of guidance from the Government about when the approach should be used and how the results may be incorporated into the planning process.

Liaison with Flood and Coastal Defence Interests

The former reluctance of local planning authorities to view erosion or flooding as anything other than a problem for the developer, has led to the necessity of managing rivers and the coastline through engineering options rather than planning control. Paradoxically, it may have been the clear success of engineered defences that has led to a lack of control of development in those areas at risk from natural hazards. For example, construction of sea defences often leads to increased pressure for development in what is now perceived to be a safe area. In reality, however, construction of sea defences only **reduces** the risk of damage. It **cannot eliminate** the risk. Increased investment and density of development behind the defences may only lead to higher losses when, inevitably, larger floods occur.

Coastal and flood defence works have been very successful in protecting communities, industrial developments or heritage sites from the threat of erosion and flooding. However, construction of the defences has had a range of effects on the physical environment; from the increase in flood risk associated with additional runoff, the encouragement of further development behind the defences to the disruption of sediment transport around the coast. Conflicts have arisen between the interests of local property owners and national conservation priorities. In some instances important habitats or geological sites are affected in order to protect property. In other instances threatened development remains unprotected because of the need to preserve the character of the unspoilt coast, (e.g. at Easton Bavents in Suffolk).

Table 13.3 Selected examples of coastal planning policies in Sefton, Merseyside (Sefton MBC, 1991).

The recognition of the dynamic nature of the coastline and its high conservation value has led to the development of policies to safeguard the coast as a natural sea defence and to avoid the need for additional engineering works:

"Consent will not normally be granted for any development proposal within the Coastal Planning Zone which would:

(i)   increase the risk of loading or coastal erosion through its impact on natural coastal processes, or

(ii)   prejudice the capacity of the coast to form a natural sea defence or adjust to changes in conditions,without risk to life or property, or

(iii)   increase the need for additional sea walls or other engineering works for coast protection purposes except where necessary to protect existing investment, or achieve the Councils strategic planning objectives". (Policy CP21).

Sefton MBC have also recognised the importance of considering coastal processes within the context of Environmental Assessment requirements:

The Council may require an environmental statement to accompany development proposals within the Coastal Planning Zone in which attention should be paid to the following matters and to any ameliorative measures required;

(i)   the coastal process of sediment transport, erosion and accretion,

(ii)   the need for coast protection and/or sea defence measure arising from the proposal, and the assessment of the effects on these measures,

(iii)   the effect of the development on the nature conservation value of the coast,

(iv)   the effect on the landscape and amenity value of the coast,

(v)   any other matters which it is necessary to consider.

Where development requires work outside the application site in order to minimise its impact on the Coastal Planning Zone, applicants may be required to enter into Section 106 agreements to ensure the implementation of such measures as:

(i)   coast protection and sea defence works

(ii)   measures to restore or enhance nature conservation or assist the site management

(iii)   provision of landscaping

(iv)   provision or replacement of access arrangements

(v)   retention, restoration or enhancement of site features and artifacts". (policy CP22)

In view of the highly sensitive nature of the coast, the council have also adopted a presumption against changes in the areas or volume of industrial sands to be extracted in the coastal zone:

"There will be a presumption against granting consent for the extraction of sand or other forms of aggregates from within the Coastal Planning Zone. The extraction of industrial sand from the Horsebank will continue under the existing consents will be subject to careful consideration". (Policy CP28).

It is clear, therefore, that there needs to be greater coordination between land use planning and strategies for flood and coastal defence, i.e. an integrated approach to management. Bearing in mind that there is essentially a presumption in favour of development proposals that are in accordance with the development plan (unless material consideration indicate otherwise) it is important that development plan policies and proposals take account of the strategic nature of many erosion, deposition and flood–related issues. In this respect local planning authorities in England and Wales have the opportunity to benefit from the major changes in flood and coastal defence that have occurred since the mid 1980's, i.e. the creation of the NRA and its policy in preparing Catchment Management Plans, the formation of Coastal Defence Groups and the development of Shoreline Management Plans for their area.

Since the late 1980's, local planning authorities in England and Wales have been increasingly influenced by the way the NRA has addressed its flood defence responsibilities. The relationship between the NRA and local planning authorities is developing into an important mechanism for ensuring that the planning system takes a broader perspective of flood–related issues than was possible in the past. Amongst the most important mechanisms for achieving this are:

(i)   the preparation of **guidance notes** for local planning authorities which address a variety of issues related to the water environment, including flood defence, (NRA, 1994);

(ii)   its role as a statutory consultee in the preparation of development plans. Circular 30/92 identified that the main input would

157

be the preparation flood risk maps (S.105 surveys; see Chapter 5), enabling the NRA to influence development patterns in a positive rather than a reactive way (DoE, 1992);

(iii)   consultation on individual planning applications where significant flood defence considerations arise. Here the NRA has two roles; considering how development would affect rivers and flood defence operations, and advising how development would increase flood risk (Circular 30/92; DoE, 1992b). A useful summary of the efficiency and effectiveness of the NRA's planning–related activity is provided by Pickup (1992);

(iv)   the preparation of **Catchment Management Plans** (CMP's) to coordinate the wide range of NRA water management functions (e.g. water quality, water resources, flood defences, fisheries, recreation, conservation and navigation). These plans are intended to provide a strategic approach to management and the means of resolving conflicts which otherwise may exist through the integrated management of river corridors, (Pickles and Woolhouse, 1994). The plans could make a positive input to the development plan process, by providing a focus for the **strategic consideration** of flood defence issues on a catchment basis. This perspective is of particular value where the NRA is required to consider:

● the effects of development on the water environment, especially the generation of additional runoff;

● the effects of flood defence works on other aspects of the water environment. For example, improved flood control may result in greater extremes of river flow, reducing aquifer recharge and affecting nature conservation interests.

Coastal Defence Groups, comprising a combination of operating authorities and other bodies with coastal responsibilities, have been established around most of the coast of England and Wales (Figure 3.2). These groups are voluntary and provide a forum for the development of Shoreline Management Plans, setting out broad objectives for the future management of the relevant shoreline (MAFF, 1994). MAFF is actively encouraging the preparation of plans for each littoral cell or sub–cell around the coast (Figure 3.3) and has indicated that they should address a wide range of management issues, including: coastal processes, planning and land use, the natural and built environment, coastal defence issues and strategic options (MAFF, 1994).

Both the NRA and Coastal Defence Groups can have a significant role in providing support to the way local planning authorities address erosion, deposition and flooding issues. Indeed, the preparation of management plans based on the operation of processes over broad physical systems, can highlight the key issues which should be taken into account by planners at a variety of scales from regional to local. NRA plans can also provide planners with a broad framework for considering the relationship between physical processes and water resource management objectives. The relationship between these plans and the planning system is, however, hindered by a lack of guidance on how local planning authorities could use the information within the plans to develop appropriate land use policies in support of broader water or coastal management objectives and the extent to which they can contribute to their preparation.

The preparation of these management plans is very much in its infancy. Their potential is clear; whether this potential is realised will depend on how effectively the broad strategies outlined in these documents are supported by land use planning policies in development plans.

Summary

Erosion, deposition and flooding processes operating on hillslopes, river systems and the coast can generate a range of issues that are relevant to land use planning. These issues are often complex, reflecting the fact that the processes are natural phenomena that are a necessary component in shaping and maintaining the nature environment; they only become hazards where development encroaches into vulnerable areas. The more important issues are related to:

● the impact of the processes on property, infrastructure and services;

● the effects of development on the degree of risk elsewhere;

- the conflicts generated by the selection of hazard management strategies.

The planning system can be used in a variety of ways to manage erosion, deposition and flooding related issues, as described in Chapters 8–12. However, it is clear that these opportunities are not fully utilised by local planning authorities, with around 50% of surveyed authorities having no planning policies addressing any aspect of these issues (Table 13.1).

The reasons why the planning system has not realised its potential as an instrument in managing these processes are, however, more a reflection of the way the system currently operates than deficiencies in the legislative provisions. Indeed, the model standards set out earlier to define a system that could effectively address erosion, deposition and flooding issues are largely achievable through the Town and Country Planning Act 1990 or the 1972 Act for Scotland.

The Government's strategy for the use of the planning system to manage these issues, in England and Wales, as defined in PPG 14, Circular 30/92 and PPG 20, concentrates on ensuring that many unsuitable areas are avoided or that appropriate precautions are taken. There are significant omission from the strategy, most notably the way the planning system considers the interactions between processes throughout catchments and coastal systems, and the complex range of indirect impacts that the processes and management responses can generate. In Scotland, the absence of planning guidance on matters relating to erosion, deposition and flooding is seen as a significant constraint on the effective operation of the planning system.

Additional factors which have limited the effectiveness of the planning system in managing physical processes include:

- the general lack of awareness of the significance of physical processes amongst many local authority planners;

- the view that physical processes are technical matters and not planning issues;

- the response of some Planning Inspectors to appeals against the refusal of planning permission on the grounds of flood risk.

To achieve the full potential of the planning system effort needs to be directed towards **informing** planners of the relevance of erosion, deposition and flooding issues and the range of management responses that can be used to achieve land use planning objectives. This need is addressed in Chapter 14 which presents a framework of advice for planners and developers on how erosion, deposition and flooding issues are best investigated and managed.

Chapter 13: References

Boothroyd J.M. 1991. The local authority requirement. In Applied Earth Science Mapping, 14–16. HMSO.

Brook D. and Marker B.R. 1987. Thematic geological mapping as an essential tool in land–use planning. In M.G. Culshaw, F.G. Bell, J.C. Cripps and M. O'Hara (eds) Planning and Engineering Geology, Geol. Soc. Eng. Geology Spec. Public No. 4, 211–214.

Brook D. and Marker B.R. 1988. Geomorphological information needed for environmental policy formulation. In J.M. Hooke (ed) Geomorphology in Environmental Planning, 247–260. John Wiley and Sons.

Department of the Environment 1979. New Ideal Homes: Land at Horton Road, Datchet. The Inspector's Report to the Secretary of State, DoE.

Department of the Environment 1990. PPG 14 Development on Unstable Land. HMSO.

Department of the Environment 1991. Applied Earth Science Mapping. Proc. Seminar at the Geological Society 21 May 1990. HMSO.

Department of the Environment 1992a. PPG20 Coastal Planning. HMSO.

Department of the Environment 1992b. Circular 30/92 Development and Flood Risk (MAFF Circular FD1/92; Welsh Office Circular 6892). HMSO.

MAFF 1994. Shoreline Management Plans: interim guidance on contents and procedures for developing shoreline management plans.

Neal J. and Parker D.J. 1988. Flood plain encroachment: a case study of Datchet, UK. Middlesex Univ. Geography and Planning Paper No. 22.

NRA 1994. Guidance notes for local planning authorities on the methods of protecting the water environment through development plans. NRA Bristol.

Pickles L. and Woolhouse C. 1994. Catchment management plans: a strategic view of flood defence. In Proc. of the MAFF Conference of River and Coastal Engineers.

Pickup M.D. 1992. Efficiency and effectiveness of planning activities. Report to NRA. Michael Parker Associates.

Rendel Geotechnics, 1995. Erosion, Deposition and Flooding in Great Britain. Methodology Report. Open File Report Held at the DoE.

Scottish Office Environment Department 1995. Planning and Flooding. National Planning Policy Guidance (draft).

Sefton MBC 1989. Coastal management plan.

Sefton MBC 1991. A Plan for Sefton.

Turner J.S. 1987. Written contribution to Discussion Session 12. In M.G. Culshaw, F.G. Bell, J.C. Cripps and M. O'Hara (eds) Planning and Engineering Geology. Geol. Soc. Eng. Geology Spec. Public. No. 4, 623–624.

Tyrrell K. 1994. A local authority approach to shoreline strategies. In Proc. of the MAFF Conference of River and Coastal Engineers.

Waveney DC 1993. Planning for the coast – a consultation document.

14 A Framework of Advice for Planners and Developers

Introduction

The operation of physical processes on hillslopes, within river channel networks and at the coast are of considerable interest to landowners, engineers, planners and land managers as well as the scientific community. Erosion processes operate to slowly break down the fabric of the land; resulting material is then transported by wind or water and deposited elsewhere in new landforms. Land can be lost; channels and reservoirs choked with sediment leading to flood problems or reduced storage capacity; new land can be created; landforms can build up to give increased protection against erosion or flooding.

Erosion, deposition and flooding are natural phenomena. The processes are an integral part of the natural landscape, especially the dynamic environments of river channels and the coast. The processes can shape the landscape forming, for example, the coastal cliffs and broad meandering rivers that are part of our natural heritage. They create and sustain valued habitats and maintain important recreational beaches or sand dunes.

The processes only become hazards or problems when society encroaches into these dynamic environments either for housing or development in, for example, floodplains or coastal cliffs or for transportation and trade along canals, rivers and estuaries. Here, attitudes to erosion, deposition and flooding can vary dramatically. To some the processes are an acceptable risk associated with living in desirable locations such as close to riverbanks or coastal cliffs; adjustments to the risk, such as flood proofing, can be made to mitigate against the effects of potentially damaging events. Others may be completely unaware of potential problems in an area; to them the sudden occurrence of an event may lead to unacceptable levels of loss.

In other instances the operation of the processes may go largely unnoticed by much of society. Erosion of riverbanks and hillslopes, for example, generally does not involve dramatic events; small amounts of material are regularly detached and carried away. The cumulative effects, however, can lead to serious consequences such as the gradual silting up of reservoirs, canals, rivers and estuaries which make regular dredging essential for maintaining navigable channels. Elsewhere, deposition can have important benefits through sustaining beaches, sand dunes, mudflats and saltmarshes; these landforms can provide natural coastal defence as well as being of conservation value.

The landscapes within which these processes occur are largely predetermined by topography, underlying geology and soils, the vegetation cover and land use. These conditions control the erosion, deposition and flood behaviour of an area or region. However, the geomorphological processes at work on the hills, rivers and coastline are not constant and do not take place in a static landscape. The occurrence of significant events is inexorably linked to the pattern of storms or severe weather conditions that are that are characteristic of Britain's maritime climate. Within this context, a number of key points are highly relevant to explaining the pattern of events:

- precipitation is usually highest in upland Britain with the result that most geomorphological energy is concentrated here. However, the uplands tend to be underlain by resistant rocks and, hence, much of the potential is often only realised in very large storms or high intensity rainfall events;

- extreme rainfall conditions can occur throughout Britain, although outside the upland areas their occurrence is infrequent. As a consequence, major events are largely unexpected in many parts of lowland Britain and, hence, their occurrence can have severe consequences;

- the rate at which the flood discharge and, hence, flood height increases with larger return periods varies across the country. For much of upland Britain, the 100 year flood event is expected to be around double the annual flood (Figure 6.2). However, in Southern England, east of a line from the Humber to the Exe, the 100 year flood event can be over 3 times the annual flood. This pattern is largely a function of catchment character and has major implications for the size of defences needed to protect against floods of a given magnitude;

- stream power, a measure of the effectiveness of river flows in achieving erosion and deposition, tends to be as much as a thousand times greater in rivers in upland areas than in lowland areas.

- the energy arriving at the coast, through waves and tides, is considerably greater than that available inland. The coastline, therefore is the most dynamic environment in Britain, although the nature of the response is controlled by the strength and disposition of materials and landforms.

- although estuaries are perceived to be quiet backwaters, they do receive extremely high energy inputs as indicated by the semi-diurnal input of over $1000Mm^3$ of water into estuaries such as the Thames, Humber and Severn. However, these forces are controlled by the development of equilibrium morphology with changes occurring twice daily and, hence, almost imperceptibly. Thus major changes in estuaries are regular and slow, involving channel adjustments over extremely long time periods in contrast to the open coast or rivers where change can be intermittent and rapid.

In certain areas physical processes, in particular river and coastal flooding and coastal erosion, can impinge **directly** on the land use planning functions of local planning authorities. This will include situations where proposals for new development or redevelopment take place in areas which may be at risk, or where proposals to protect

areas of land from physical processes can have an impact on other planning objectives.

Erosion, deposition and flooding can also generate **indirect** land use issues. Bank erosion, together with sediment delivered to watercourses from hillslopes, can lead to sedimentation problems and increased flood risk; the response, maintenance dredging, can cause land use issues when the material has to be disposed of on land or if the operations threaten to damage conservation interests. Conversely, the cessation of dredging following the closure of port and harbour facilities may lead to the degradation of the riverbank environment in city centres or cause an increase on flood risk. These matters can clearly be considered as planning issues and may need to be addressed by local planning authorities.

Erosion, deposition and flooding processes and the way they are managed can have wide ranging effects on the water environment, especially within river systems and on the coast. Of particular importance is the dependency of many nature and geological conservation sites, fisheries spawning areas and recreation interests on the uninterrupted operation of the processes; erosion or flood control works can lead to degradation of the natural environment and conflict with these interests. In other instances the exploitation of mineral resources may be constrained by the need to ensure that extraction operations or subsequent land restoration does not affect the level of risk to developments elsewhere.

The Degree of Risk

In some areas erosion, deposition and flooding events can be annual occurrences, as on many lowland rivers (**hence the term floodplain**) or unprotected soft rock cliffs. In other instances events may be associated with relatively infrequent events, such as high intensity rainstorms or storm surges on the coast, or the failure of man-made structures such as reservoirs or embankments.

Disasters and other damaging events occur because development has taken place in areas where there are risks due to the nature of the topographic and geological setting. The degree of risk can be expressed in a variety of ways. The most common is to use the **return period** or **recurrence interval** of an event of a particular size. There are, however, frequent problems in the public perception of this approach as events tend to be random and not regular in occurrence. Thus, the 100 year flood has an **average** return period of one century, but could occur any year. Indeed, the likelihood of such a 100 year flood event occurring during the lifetime of a building (taken here as 60 years) is around a 45% chance (Table 4.7).

A different perspective is provided by considering the degree of risk in terms of the standard of protection which exists in an area. MAFF, for example, have set out **indicative standards of protection** in terms of flood return periods for five subjectively expressed current land use bands (Table 11.7); these standards are intended as guidance, not to set minimum standards of protection. Similarly large raised reservoirs are currently classed in four hazard categories, mainly according to the level of development downstream. The standard of safety required in an agricultural area would be related to a 150 year flood, whereas a reservoir where a breach will endanger lives in a community would be designed to cope with a 10,000 year flood (which has less than a 0.1% chance of occurring in any year).

Construction of river and coastal engineering works only **reduces the risk of damage; it cannot eliminate the risk**. Increased investment and density of development behind the defences may lead to higher losses when, inevitably, larger events occur and to more pressure for expensive maintenance and replacement of defences as they become worn. A balance has to be found between the costs of providing defences and the benefits to the nation as a whole. To attempt to protect all rivers or the entire coastline would not only be uneconomic but could intensify the problems in many areas. For example:

- construction of flood embankments reduces floodplain storage and can lead to more severe flooding downstream;

- river channelisation may lead to erosion and deposition problems both upstream and downstream of the works;

- coastal defences may disrupt the supply or transport of sediment around the coast and have an adverse effect on defences elsewhere.

Flood and coastal defence works may also have significant consequences for other river and coastal interests, especially areas of nature or geological conservation value.

Management of Erosion, Deposition and Flooding Processes

Management of erosion, deposition and flooding processes is dominated by the need to reconcile a number of conflicting demands:

- protecting vulnerable communities, important economic resources and facilities;
- facilitating the provision of competitive ports and navigable waterways;
- meeting the demands of the rapidly-expanding tourism and recreation industries;
- protecting areas of scenic, geological or ecological importance.
- protecting the marine environment.

Finding the right balance may lead to disputes. Some existing communities may feel that conservation interests are creating unnecessary levels of risk to property in vulnerable areas, whilst others may feel that areas of national importance have been progressively spoilt by inappropriate development and the subsequent need for flood or coastal defences. Acceptable solutions to these conflicts can only be reached by ensuring that a wide range of interests are considered on the decision-making process.

A variety of responses can be practised in areas prone to erosion, deposition and flooding, including **acceptance** of the risks, **avoiding** vulnerable areas, **reducing** the occurrence of potentially damaging events and **protecting** against such events. In most cases the response will be complex, involving a variety of measures adopted by residents, landowners and different authorities at different locations within a catchment or coastal system. The most appropriate option will depend on the nature of the problems, the level of acceptable risk, the availability of resources and the statutory powers available to interested bodies or authorities to address the issues.

Precautionary measures needed to reduce the levels of risk may, however, have significant and irreversible effects on some environmental resources. Judgements have to be made about the weight to be put on those factors in particular cases. Sometimes the environmental costs may have to be accepted as the price of economic development, but on other occasions resources may be so valuable that they have to be protected from the potential effects of development. It is important, therefore, that management decisions are based on the best possible understanding of the physical environment and how to exploit, manage and protect it. This can only be achieved through an awareness of the operation of catchments and coastal systems.

The primary responsibility for dealing with erosion, deposition and flooding problems lies with the landowner. However, where it has been considered unreasonable to expect a landowner to be responsible for protection of vulnerable areas, statute law allows various authorities to intervene and provide defences. **Flood and coastal defences** can be provided by various operating authorities where such works are seen to be in the national interest. The powers contained in the Acts listed below are **permissive** in that they allow works to be carried out, but do not require them (ie. they are not mandatory powers):

- **Coast Protection Act** 1949
- **Flood Prevention (Scotland) Act** 1961
- **Water Resources Act** 1991 (England and Wales)
- **Land Drainage Act** 1991 (as amended by the **Land Drainage Act 1994**; England and Wales)

Schemes are promoted by operating authorities (the NRA, Internal Drainage Boards, local authorities in England and Wales; Regional Councils in Scotland) and may be grant–aided by the relevant Government department (MAFF, the Welsh Office or the Scottish Office). In England and Wales, schemes are evaluated on a combination of technical, economic and environmental factors, as outlined in the recent Strategy for Flood and Coastal Defence (MAFF/Welsh Office 1993).

In many instances conservation status has been used to prevent development in some areas of potential problems. Many undeveloped coastlines, for example, are protected by designation as part of National Parks or AONB's (National Scenic Areas in Scotland) and their definition as heritage coasts. Rivers and the coast also provide natural habitats and geological features of national and international importance which have been protected as SSSI's, National Nature Reserves, Ramsar Sites, Special Protection Areas, Special Areas of Conservation or UNESCO Biosphere Reserves. The management of these conservation resources can benefit from an awareness of the extent to which the features are dependent on continued erosion, deposition and flooding for maintaining their conservation value and their sensitivity to the effects of development. A variety of management plans are being

developed to provide a mechanism for reconciling flood and coastal defence interests with the need to conserve and enhance the natural environment. These include: Catchment Management Plans, Water Level Management Plans and Shoreline Management Plans.

The **planning system** can play an important role in the way erosion, deposition and flooding issues are managed. This may involve minimising the risks and effects of the processes on property, infrastructure and the public, ensuring that development is not placed in unsuitable locations without adequate precautions, taking account of the possible adverse effects these precautions may have on the environment, ensuring that development does not interfere with channel maintenance operations and protecting areas of conservation value and supporting management plans. A local planning authority does not owe a duty of care to individual landowners when granting applications for planning permission and accordingly is not liable for losses arising as a result of development having been permitted. However, where development is proposed on land known to be at risk, local planning authorities have a duty to take account of all material considerations, including the various issues related to erosion, deposition and flooding, before reaching a decision on whether to grant planning permission.

The Role of the Developer

The responsibility for determining whether land is suitable for a particular purpose rests primarily with the developer (Figure 14.1). The developer should therefore make a thorough investigation and assessment of the site and surrounding area to establish whether:

- the land is capable of supporting the load to be imposed;

- the development will be threatened by erosion, deposition or flooding problems;

- the development will affect the level of risk from erosion, deposition and flooding problems to property, infrastructure and services in adjacent areas or elsewhere within a catchment or coastal system.

The assessment of erosion, deposition and flooding problems and the associated risk requires careful professional judgement. Developers should seek

Figure 14.1 The role of the developer in the investigation of sites.

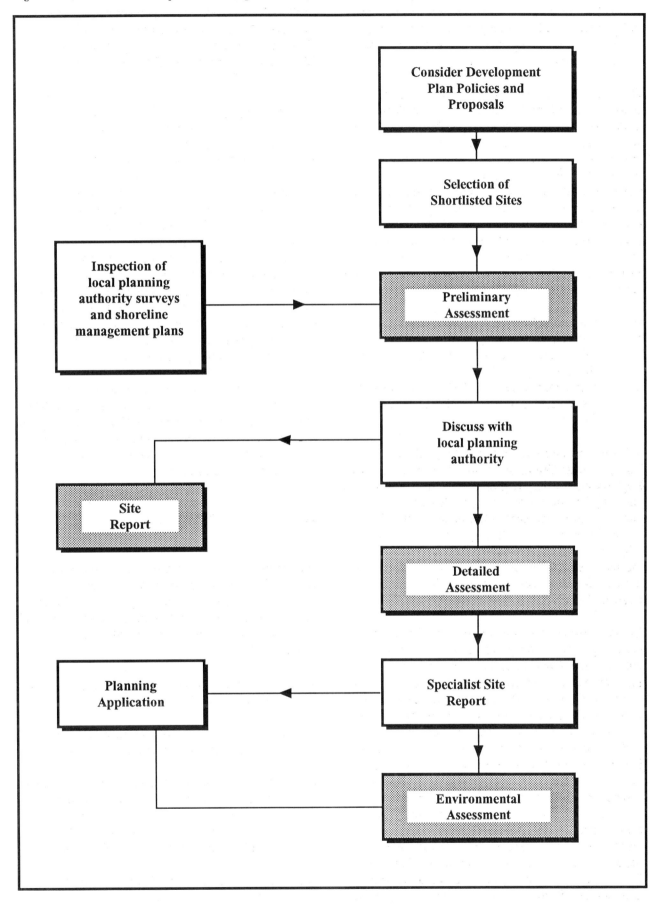

expert advice about the likely consequences of proposed developments on the physical environment. This advice will generally involve some form of **investigation** into the nature of the problem and should provide an indication as to whether the site is suitable or whether precautionary measures are needed to prevent problems affecting the site or neighbouring areas. **Site reconnaissance** (preliminary assessment) should be carried out to determine whether the proposed sites are in vulnerable locations and, if so, the likelihood of potentially damaging events. The need for new defences or improvements to existing defences should be identified, assessing whether such works are likely to be cost effective or environmentally acceptable.

The first stage in any preliminary assessment should be a **desk study** of available records, maps and reports. This study should concentrate on a range of factors relating to the ground conditions at the site or shortlisted sites. Although the study should be focused on the conditions at the relevant sites, it will also need to consider the broader setting to establish whether processes operating on adjacent land could affect the site, or whether development of the site could lead to problems elsewhere.

The desk study should be accompanied by some form of **aerial photograph interpretation and ground inspection** to confirm or determine the nature of the conditions at the different sites. This inspection should complement the desk study and allow **preliminary assessments** of **risk** to be made, together with the potential environmental effects. From this a judgement may then be made about whether the site or sites require defence works and, if so, the likely scale and costs of those works.

The developer should provide sufficient information to enable to local authority to determine the planning application. Indeed, the authority is entitled to require the developer to seek suitable expert advice. Advice on the level of information which needs to be presented with planning applications should be sought from the local planning authority at an early stage. This is likely to vary according to the severity of the potential problems and the nature of the proposed development, and may involve:

- a **general requirement** to satisfy the authority that the site is not at risk, that any potential problems can be overcome and the development will not have

significant effects on other interests. In many instance this will involve a **desk study** and some form of **ground inspection**;

- a requirement for **specialist detailed investigations** where potential problems are known or suspected to exist, addressing the relevant issues and indicating how they would be overcome without affecting other interests. This would typically involve a combination of desk study, ground inspection, field measurement, sub–surface investigation, laboratory testing and computer modelling;

- **environmental assessment** (EA) where the development is considered by the local planning authority to have the potential for causing significant environmental effects.

If the developer's **site report** (see below) indicates that erosion, deposition and flooding problems can be avoided or accommodated planning permission may be granted, unless the application fails to meet other planning criteria. In some cases, planning permission may be conditional on the incorporation of any precautionary or remedial measures (recommended in the site report) in the detailed design.

The Role of the Local Planning Authority

There are considerable opportunities to prevent or reduce damage to new development through the consideration of erosion, deposition and flooding at all stages of the planning framework from regional to local levels (Figures 14.2 and 14.3). Local planning authorities should also liaise with catchment and shoreline managers in developing a coordinated approach to managing erosion, deposition and flood related issues (see below).

Natural processes operate over a much broader scale than individual authority areas. As a result land use policies and proposals in one area may have an adverse effect downstream or on neighbouring sections of the coast. **Regional planning groups** should, therefore, liaise with catchment managers (e.g. the NRA) and Coastal Defence Groups to identify those erosion, deposition and flood related issues that need to be considered within the context of a catchment or coastal system. Possible examples include the effect of development on the disruption of the

Figure 14.2 Forward Planning: the need for earth science information.

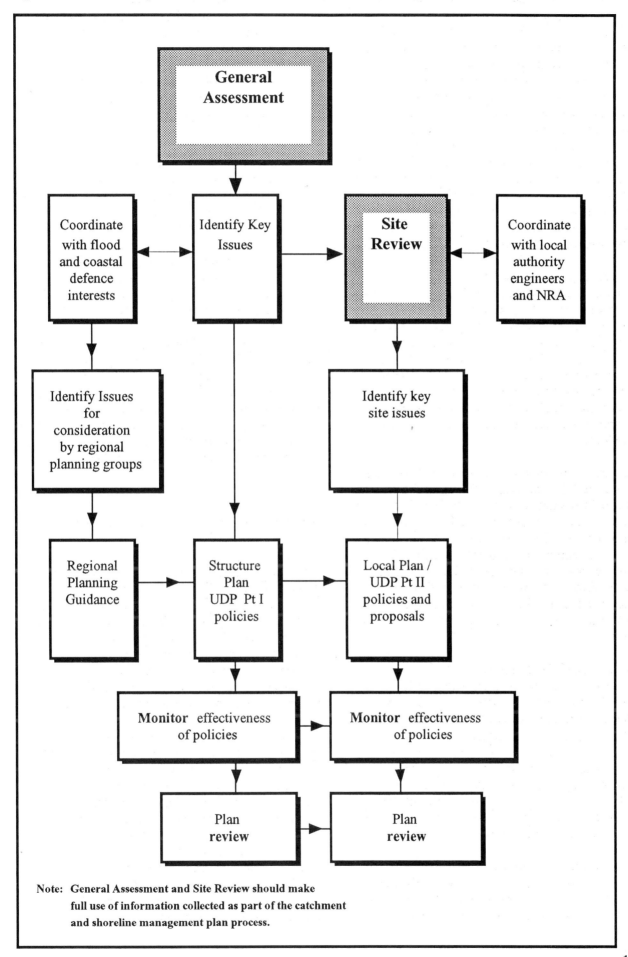

Note: General Assessment and Site Review should make
full use of information collected as part of the catchment
and shoreline management plan process.

167

Figure 14.3 Control of development: the need for earth science information.

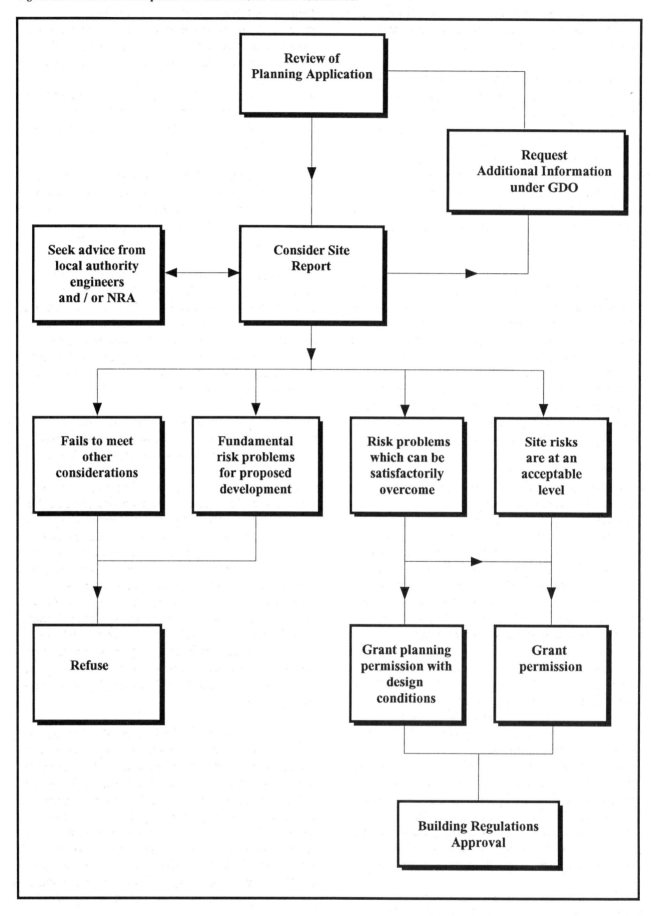

natural movement of coastal sediments and the impact on natural coastal defences and conservation interests, and the reduction of floodplain storage within catchments.

Structure Plans or **Unitary Development Plans (Part I)** provide an excellent opportunity for identifying the extent of erosion, deposition and flooding problems in an area and outlining the range of policies that are to be adopted in these areas. The scale of provision to be made to housing, employment opportunities, etc. in the area as a whole and the breakdown for each district should take account of any erosion, deposition and flood related issues, including the need for flood and coastal defences, and the protection of conservation interests.

The **Local Development Plan** or **Unitary Development Plan (Part II)** can be used to set out detailed policies to address these issues, the basis for determining planning applications and the types of planning conditions normally expected to be met. When evaluating potential sites for development or redevelopment, the local planning authority should consider whether flood or coastal defences, if necessary, could be undertaken without adversely affecting other interests.

Local Plans should define those areas where a particular consideration of the implications of erosion, deposition and flooding is necessary, either in narrative form or by reference to maps of resources and constraints (see below). It should be made clear that these maps have been compiled from the most reliable information available and that they are not necessarily definitive statements on the nature and extent of potential problems in an area; the responsibility for determining whether a particular site is suitable lies with the developer.

The handling of planning applications in areas prone to erosion, deposition and flooding will need to take full account of the potential problems that these processes could cause to the proposed development, the neighbouring area and elsewhere within a catchment or coastal system. It is important to stress the value of prior consultation between a developer and the local planning authority before an application is submitted and the need for the developer to provide sufficient information to enable the authority to consider the application, after appropriate discussions with statutory consultees and other interested bodies.

The authority should then determine whether a proposed development should proceed, taking into account **all relevant material considerations**. A development may be approved subject to **conditions** specifying measures to be carried out in order to minimise risk or reduce the effects of the development on other interests. These may include:

- building design; minimum floor heights or means of escape in flood risk areas, flexible constructions on unstable coastal slopes;

- occupancy; limiting periods of occupancy in flood risk areas to periods when flood risk is significantly reduced or controlling the use of basements for apartments in vulnerable areas.

If development is approved, it is then the local authority's responsibility under the **Building Regulations** to determine whether the structure can be built and used safely. Even though the local authority may have granted planning permission, the responsibility and subsequent liability for safe development and secure occupancy rests with the developer and/or landowner.

Some development is deemed not to require planning permission, and is, therefore **permitted development**. Such developments may relate to matters which lie outside the terms of the planning legislation, such as agricultural buildings, or very small–scale ("de minimis") operations such as the construction of swimming pools, terracing of gardens, removal of vegetation from slopes and building improvements. These may not be of concern in most locations but can all contribute to coastal cliff erosion problems. In some areas, therefore, the local authorities might give serious consideration to making a **direction under Article 4 of the General Development Order, 1988** to remove specific development rights within an area and require planning permission to be obtained. In other cases, an advisory code of good practice may be appropriate.

The UK is bound by **EC Directive 85/337/EEC** on the "assessment of the effects of certain public and private projects on the environment". This directive requires an **environmental assessment** (EA) to be carried out before development consent is granted for certain types of major project, listed in two Annexes to the directive. For Annex I projects EA is mandatory. For Annex II projects EA is required if there are likely to be significant environmental effects. Where EA is required, the developer must prepare and submit an **environmental statement** setting out their own assessment of the likely environmental effects of the proposed development.

The planning system is one of the main instruments for taking account of an EA under the **Town and Country Planning (Assessment of Environmental Effects) Regulations 1988.** The regulations apply to certain projects that require planning permission and included within the Annex II schedule are: canalization or flood relief works; dams and reservoirs and coast protection works.

In deciding whether an EA is required, local planning authorities should bear in mind the potential effects of these projects on the level of risks elsewhere, local amenities and conservation interests. EA may be particularly important for projects:

- in particularly sensitive or vulnerable locations;

- which could give rise to particularly complex adverse effects, for example, in terms of disruption in the transport of coastal sediment or changes in the flow regime of a river.

Planning Policies and Proposals

Structure Plans and Unitary Development Plans (Part I) should set out the broad strategy to be adopted in an area for addressing erosion, deposition and flooding issues. This strategy should recognise the need to avoid placing additional development in vulnerable areas, thereby increasing the number of areas requiring defence works. It is important that these strategic policies are developed in consultation with neighbouring local planning authorities and the relevant operating authorities. They should be consistent with the flood and coastal defence policies set out in Catchment Management Plans and Shoreline Management Plans, and take account of relevant Regional Planning Guidance.

The detailed policies and proposals presented in Local Plans and Unitary Development Plans (Part II) should be consistent with the broad strategy defined by the Structure Plan or Unitary Development Plan (Part I). The precise nature of the relevant policies will reflect the types of issues that arise in an area, but may include:

- avoiding locating new development in unsuitable areas or specifying restrictions on residential occupancy in risk areas;

- ensuring that development does not adversely affect the management of sites or areas of national and international conservation value;

- ensuring that development does not restrict access to watercourses for channel maintenance works;

- ensuring that precautions are taken to prevent runoff from new development leading to an increase in flood risk;

- ensuring that development does not affect coastal cliff stability or lead to an increase in coastal erosion.

- ensuring that flood and coastal defence works are compatible with land use planning and conservation objectives in an area.

It may be appropriate for local planning authorities to consider a presumption against development in unprotected floodplain areas and coastal lowlands, and along rapidly eroding coastal cliffs. In considering proposals for development in and around the margins of upland areas, local planning authorities will need to take account of the possible implications of permitting the development on reservoir safety standards should there be reservoirs upstream. Such developments may lead to considerable expenditure by reservoir owners or operators when safety improvements are required. Consideration may also need to be given to discontinuing or restricting existing land uses or buildings in vulnerable areas through discontinuance orders or compulsory purchase. In some vulnerable areas, conservation may be the most appropriate land use option.

Local plans should not allocate land for development on land at risk which is not currently protected, unless the developer is willing to protect the land to the appropriate standard as part of the development. Development proposals should, therefore, take due account of whether any necessary flood and coastal defence works or improvements can be undertaken by the developer without adversely affecting other interests such as conservation, tourism and recreation, navigation and fisheries. Landowners, prospective purchasers and developers will thus be forewarned of any likely problems and can estimate the cost implications. It should be noted that developers will also require the consent of the NRA (in England and Wales) and the relevant coast protection authority before commencing any flood defence or coast protection works, respectively. Defence works below High Water Mark will also require consent from:

- the relevant fisheries department (MAFF or SOAFD) for placing materials below HWM, under the Food and Environment Protection Act 1985 Part II.

- the Secretary of State for Transport to ensure that works in tidal waters do not affect navigation, under the Coast Protection Act 1949.

Information Requirements for Planning and Development

Planning involves reconciling development requirements with the need to protect, conserve and, were appropriate, improve the landscape, environment and recreational opportunities. To achieve these aims the planning system needs to be supported by an appropriate level of information on the social, economic and physical character of an area. This need for relevant information is defined by the planning legislation which places a requirement to collect information on both local planning authorities and developers.

As part of their forward planning functions, **local planning authorities** are required to keep all matters under review that are expected to affect the development of their area or the planning of its development (1972 Act S.4; 1990 Act Part II). The matters to be examined and kept under review include:

- the principal physical and economic characteristics of their area and, where appropriate, neighbouring areas;

- any changes in the physical and economic characteristics and the effect which those changes are likely to have on the planning and development of an area.

The local planning authority is entitled to require the developer, at his expense, to provide suitable expert advice and relevant information with a planning application, and is entitled to rely on that

advice in determining the application and formulating any necessary conditions. However, it is important that development plan preparation is supported by an appropriate level of information on the nature of physical conditions in an area. Failure to recognise potential at an early stage of the decision making process can lead to increased development costs, abandoned development, calls for publicly funded defence works, damage to adjacent land property and the environment, or an increase in the risks to other land users. Policy omissions can result in development proceeding in areas which might not be suitable.

General Assessments

For many purposes, such as detailed policy formulation and preparing a regional context for strategic planning, a broad brush approach which highlights the physical conditions and related issues which need to be borne in mind can contribute significantly to the safe, cost-effective development and use of land. **General assessment** of an area can be undertaken as part of the survey of the principal characteristics of the authority area in advance of a development plan review (Figure 14.4). The approach can provide a relatively quick appraisal of conditions over large areas by collecting and interpreting readily available data sources. Such studies should highlight:

- where erosion, deposition and flooding issues need to be considered;

- the nature of the hazards in the areas and the general level of risk to existing land and uses and development;

- whether potentially damaging events can be expected during the normal lifetime of particular types of development;

- the types of uses or development that, in general terms, may be best suited to those areas;

- whether development or land use can have adverse effects on other interests in the river or coastal environments, such as conservation and recreation.

It is important to stress that the assessment of erosion, deposition and flooding issues requires specialist professional judgement. In many circumstances the local authority should consult

Figure 14.4 The role of general assessment and site review in the forward planning process.

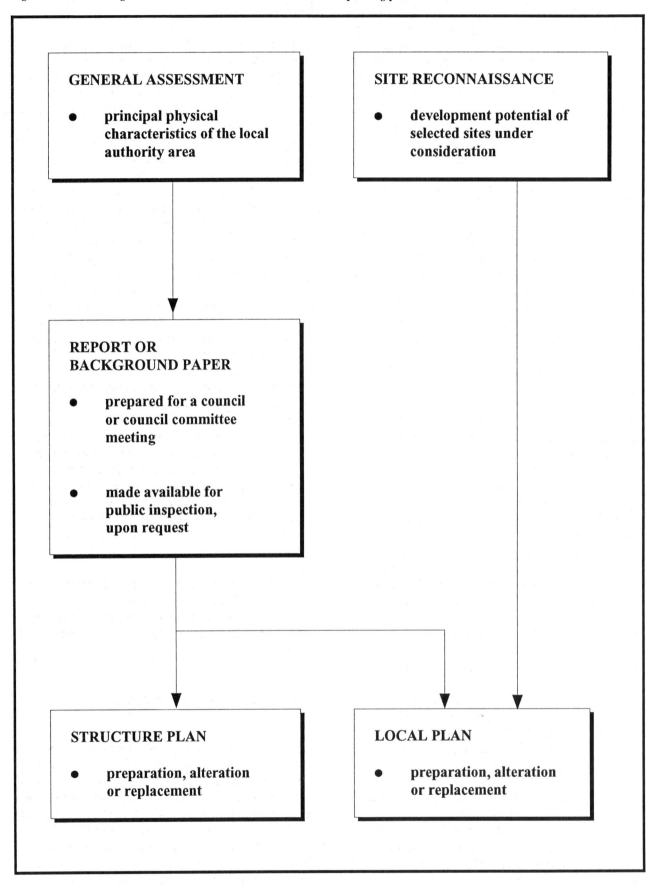

with those bodies with an interest or responsibility in these matters who may be able to supply much of the technical information required for a general assessment. Local planning authorities should make full use of the experience and expertise of other departments within the local authority, especially the engineers responsible for aspects of flood defence and coast protection. However, some local authorities may not have the required expertise available to them and it may be appropriate to consider the need to use commercial consultants.

General assessments should be essentially **desk studies**, compiling and analysing readily available information, supplemented by inspection of historical maps or aerial photographs and limited field inspection, where necessary. It will not be feasible to collect and collate all existing data; to do so may be counter–productive as the need is for a broad brush appreciation of the physical character of an area and not detailed information about well researched sites. Most investigations should, however, follow the 5 simple stages outlined below and summarised in the checklist provided in Appendix A.

Stage 1; Lack of awareness of the potential for damaging events in an area is often a significant obstacle to be overcome, especially where the threat is associated with extremely rare events (e.g. flash flooding in dry valleys and tributary streams) or failure of man–made structures (e.g. dams and embankments). It is important, therefore, that the first step in any general assessment should be to recognise the problems that can be expected to occur in an administrative area; the characteristic vulnerable settings shown in Table 14.1 with reference to particular terrain units can help authorities to identify what processes can be expected in particular landscapes (Figure 14.5).

A number of key issues are generally associated with the operation of the processes within particular terrain units. These issues can be sub-divided into 3 broad groups: **risks, sediment budget, and sensitivity**; Table 14.2 provides a checklist of issues that may need to be considered in different environments.

Stage 2; identification of key information sources associated with the various issues relevant to a particular area. Table 14.3 provides a checklist of some of these key sources and highlights the importance of accessing Catchment Management Plans, Water Level Management Plans and Shoreline Management Plans and discussing technical aspects with the NRA, coastal defence groups and coast protection authorities responsible for their preparation.

Stage 3; the degree of risk in vulnerable areas needs to be assessed in order to provide a firm basis for development plan policies and proposals. In England and Wales, **flood risk** information should be provided to local planning authorities by the NRA. On coastal cliffs, it may be necessary to define the area likely to be affected by erosion during the lifetime of a building. The episodic nature of cliff recession can, however, restrict the reliability of predictions made from average annual rates identified from historical sources. As a result MAFF have recently commissioned a research programme to develop techniques to help coastal engineers accurately and reliably predict future erosion rates on soft rock cliffs in order to assist the preparation of Shoreline Management Plans (Rendel Geotechnics, in prep). Local planning authorities should, therefore, seek the advice of coast protection authorities when considering the risks associated with cliff recession. Where "set back" lines are to be shown in development plans, they should be developed in conjunction with the relevant coast protection authority.

Stage 4; a review of the environmental resources that are dependant on the continued operation of erosion, deposition and flooding processes. This should involve consultation with the relevant conservation agencies and will need to consider regional **sediment budgets** and to identify **sensitive** elements of the landscape that may be vulnerable to the effects of development.

Stage 5; accessible presentation of information in formats suitable for use by planners and developers. This may involve using the **Applied Earth Science Mapping** approach described below.

Applied Earth Science Mapping

In the past, the planning system has not taken sufficient account of the dynamic nature of river and coastal environments. This is due, in part, to a lack of awareness of the physical environment and the limited use of technical information in support of decision making. Access to technical earth science information will often be limited to site investigation reports prepared by a developer in support of a planning application. In many instances such proposals will be considered in relation to planning policies developed with little

Table 14.1 Characteristic vulnerable settings associated with different terrain units.

TERRAIN UNIT	CHARACTERISTIC PROCESSES	VULNERABLE SETTINGS
Mountainous and Upland areas	Upland soil erosion	Upland peaty soils. Areas of recreation pressure. Recently afforested land.
	Flash floods	Land adjacent to streams. Areas downstream of dams and reservoirs.
	Sedimentation	Upland lakes and reservoirs.
Undulating lowlands	Water erosion	Areas of poor land management associated with silty and fine sandy soils.
	Wind erosion	Areas of poor land management associated with fine sandy soils.
	Flash floods	Land adjacent to tributary streams, especially around the margins of upland areas. Areas with inadequate storm drainage systems. Areas downstream of dams and reservoirs.
	Unstable river channels	Alluvial rivers around the margins of upland areas. Alluvial rivers upstream and downstream of channelisation, river engineering works, bridges etc.
	Mudfloods	Dry valleys in areas of poor land management.
	Lowland floods	Low lying land adjacent to rivers and streams.
	Sedimentation	Within alluvial river channels.
Estuaries	Flooding	Low lying land adjacent to rivers and streams.
	Sedimentation	Estuaries with a flood tide dominance which results in a net movement of sediment into the estuary.
	Channel erosion	Estuary margins developed in soft cohesive materials.
Coastal lowlands	Flooding	Low lying areas adjacent to rivers and streams. Low lying coastal land, including areas behind sand dunes, shingle ridges etc.
	Wind erosion	Areas of poor land management associated with lowland peat soils.
	Sedimentation	Within alluvial river channels
Coastal cliffs	Landsliding and cliff recession	Unprotected soft rock cliffs, especially on exposed coasts.
Sand dunes	Wind erosion	Unstabilised, bare sand areas.
	Flooding	Low lying coastal land behind sand dunes

or no regard for the physical processes and, as a result, planners without a background awareness of the significance of these processes may not be aware that the proposal could have considerable effects on property or environmental interests elsewhere within a catchment or coastal system.

It has been recognised that one of the main problems facing land use planning is that few planners ave an earth science background and few earth scientists have a planning background, and

hence there is often a communication gap between the two groups. However, both groups do share a common skill: **familiarity with maps**.

The shared background in geographical skills provides an opportunity for presenting technical information in a format that is readily appreciated by planners. However, such maps need to address planning concerns and not seek to present only specialist earth science themes. Maps for planners need to be:

Figure 14.5 Characteristic erosion, deposition and flooding problems in different terrain units.

Terrain Unit	Upland erosion	Water erosion	Wind erosion	River channel instability	Mud floods	Flash floods	Lowland floods	Coastal floods	Sedimentation	Landsliding and cliff recession	Wind blown sand
Mountainous and Upland Areas	•	•		•		•			•		
Undulating Lowlands		•	•	•	•	•	•		•		
Estuaries							•	•	•		
Coastal Lowlands			•					•	•		
Coastal Cliffs										•	
Sand Dunes											•

Table 14.2 A selection of the key issues associated with erosion, deposition and flooding in different terrain units.

TERRAIN UNIT	RISK ISSUES	SEDIMENT BUDGET ISSUES	SENSITIVITY ISSUES
Mountainous and Upland Areas	• Upland soil erosion • Flash floods • Sedimentation	• Loss of water storage capacity in reservoirs • Sedimentation in watercourses, increased flood risk and navigation problems	• Land degradation and loss of amenity • Creation of conservation sites in flood events
Undulating lowlands	• Soil erosion (water & wind) • Flash floods & mudfloods • Lowland floods • Unstable river channels • River erosion & deposition	• Sedimentation in watercourses, increased flood risk and navigation problems • Disposal of dredgings on land • Effect of mineral extraction from channels on erosion and risk • Effect of mineral stockpiles on floodplain storage	• Creation and maintenance of conservation sites • Effect of channel maintenance and flood defence on conservation sites
Estuaries	• Flooding • Sedimentation • Channel erosion	• Sedimentation in watercourses, increased flood risk and navigation problems • Disposal of dredgings on land or at sea • Maintenance of natural coast defences eg mudflats and saltmarshes • Effect of mineral extraction from channels on erosion and flood risk • Effect of mineral stockpiles on floodplain storage	• Creation and maintenance of conservation sites • Effect of channel maintenance, flood defence and mineral extraction on conservation sites • Effect of sea level rise on flood risk
Coastal lowlands	• Flooding	• Maintenance of natural coastal defences eg mudflats, saltmarshes, beaches, sand dunes • Maintenance of amenity beaches and sand dunes • Effects of mineral extraction from the foreshore on erosion and flood risk	• Creation and maintenance of conservation sites • Effects of coast protection on conservation sites • Effects of sea level rise on flood risk
Coastal cliffs	• Landsliding • Cliff recession	• Supply of sediment to natural coastal defences eg mudflats, saltmarshes, beaches, sand dunes • Supply of sediment to amenity beaches and sand dunes • Effects of mineral extraction on coastal slope stability	• Creation and maintenance of conservation sites • Effects of coast protection on conservation sites • Effects of sea–level rise on erosion rates
Sand Dunes	• Wind erosion • Flooding	• Supply of sediment to and maintenance of natural coastal defences eg beaches • Supply of sediment to and maintenance of amenity beaches and sand dunes • Effects of mineral extraction on erosion and flood risk	• Creation and maintenance of conservation sites • Effects of coast protection and mineral extraction on conservation sites • Effects of sea–level rise on erosion rates

• concise and clear summaries of the key earth science information as it relates to key planning issues;

• highlight potential problems so that users are aware of the factors which may restrict development opportunities in an area;

• indicate the types of planning response that may be appropriate to take account of particular physical conditions.

Recognition of the difficulties in incorporating earth science information into the planning process has led the DoE to actively promote the development of Applied Earth Science Mapping as a vehicle for presenting technical information to the planning community. The techniques involve collecting and collating available earth science information from various sources and then summarising it in a form specifically tailored to meet local planning needs. In general, this type of study is directed towards preparing a combination of thematic maps at a general scale, which become increasingly focused on key planning issues as they

Table 14.3 Key information sources associated with various erosion, deposition and flooding issues in different terrain units.

TERRAIN UNIT	KEY ISSUE	KEY INFORMATION SOURCES
Mountainous and Upland Areas	Risks	• Soil Survey and Land Use Research Centre soil maps and memoirs for England and Wales. • Macaulay Land Use Research Institute soil maps and memoirs for Scotland. • NRA surveys of flood risk areas.
	Sediment Budget	• NRA Catchment Management Plans.
	Sensitivity	• Conservation agency maps and records. • NRA Catchment Management Plans.
Undulating Lowlands	Risks	• Soil Survey and Land Use Research Centre soil maps and memoirs for England and Wales. • Macaulay Land Use Research Institute soil maps and memoirs for Scotland. • NRA surveys of flood risk areas.
	Sediment Budget	• NRA Catchment Management Plans. • British Geological Survey maps, reports and records.
	Sensitivity	• Conservation agency maps and records. • NRA Catchment Management Plans.
Estuaries	Risks	• NRA surveys of flood risk areas.
	Sediment Budget	• NRA Catchment Management Plans. • Estuary Management Plans. • Shoreline Management Plans. • Water Level Management Plans. • British Geological Survey maps, reports and records. • Macro-review of the Coastline (HR, Wallingford, various dates).
	Sensitivity	• JNCC Directory of the North Sea Coastal Margin (Doody et al, 1993) • JNCC Inventory of UK estuaries (Buck, 1993).
Coastal Lowlands	Risks	• NRA surveys of flood risk areas. • Shoreline Management Plans.
	Sediment Budget	• Shoreline Management Plans. • British Geological Survey maps, reports and records. • Macro-Review of the Coastline (HR Wallingford, various dates).
	Sensitivity	• Conservation agency maps and records. • Shoreline Management Plans. • Coastal Management Plans. • JNCC Directory of the North Sea Coastal Margin (Doody et al, 1993).
Coastal Cliffs	Risks	• National Landslides Databank (Rendel Geotechnics). • British Geological Survey maps, reports and records. • Aerial Photographs. • Historical topographic maps. • Shoreline Management Plans.
	Sediment Budget	• Shoreline Management Plans. • British Geological Survey maps, reports and records. • Macro-Review of the Coastline (HR Wallingford, various dates).
	Sensitivity	• Conservation agency maps and records. • Shoreline Management Plans. • Coastal Management Plans. • JNCC Directory of the North Sea Coastal Margin (Doody et al, 1993).
Sand Dunes	Risks	• Shoreline Management Plans.
	Sediment Budget	• Shoreline Management Plans. • British Geological Survey maps, reports and records. • Macro-Review of the Coastline (HR Wallingford, various dates). • Mapping of Littoral Cells (HR Wallingford, 1993). • Beaches of Scotland (Ritchie and Mather, 1985).
	Sensitivity	• Conservation agency maps and records. • Shoreline Management Plans. • Coastal Management Plans. • JNCC Sand Dune Vegetation Survey of Great Britain (Dargie 1993). • JNCC Directory of the North Sea Coastal Margin (Doody et al, 1993).

are developed from the basic factual information. For this reason, Applied Earth Science Mapping studies have comprised:

i   a set of **element maps** depicting factual information on specific geologically related topics (e.g. rock types, geomorphology, soils, slope steepness, mineral workings, made ground, etc);

ii   a set of **derivative maps** which draw on the basic data to define characteristics of particular interest such as possible mineral resources, geotechnical conditions, landslip potential, etc; and

iii   a set of **summary maps** compiled from the element and derivative maps which highlight the general characteristics of an area in terms of **resources for development** and **constraints on development**.

These maps should provide a snapshot of possible limitations and advantages in an area which could be directly set against other factors which are already considered by the planner, such as existing and proposed land uses, conservation and amenity areas, agricultural land potential, communications and natural habitats. The aim is not that earth science information should necessarily over–ride all other factors but that it should be capable of being considered alongside those other factors and given a weighting appropriate to the particular circumstances.

Applied earth science mapping studies should be presented at around 1:50,000 to 1:10,000 scale. A wide range of maps can be custom made on specific topics and for specific audiences. By and large planners are served by summary and derived maps which allow them to devise broad, strategic policies for development and conservation and indicate, in general terms, areas in which certain types of development might be preferred or opposed. Their technical advisors will generally find the greater detail of the element and derivative maps of the more use.

Planning Guidance

There are considerable opportunities to prevent future damages to property or environmental degradation through incorporating the knowledge of risks, sediment budget and sensitivity within the existing planning framework. This information can be used to form the basis of **planning strategies** which reflect variations in physical conditions rather than adopting a blanket approach to planning. Knowledge of physical conditions can be summarised as preliminary planning guidance which indicates how different areas need to be treated for both policy formulation and development control. Areas may be recognised which are likely to be suitable for development, along with areas which are either subject to significant constraints or mostly unsuitable. Advice can also be presented on the level of supporting earth science information which should be presented with planning applications in different areas. It is important to note that there may be areas where there is insufficient background information on which to develop planning strategies; in such circumstances, it may be necessary to adopt a precautionary approach whereby development would only be approved after the developer as been able to fully demonstrate that the proposed site can be safely developed, without adverse effects elsewhere. Table 14.4 summarises the general guidance that has been prepared for a selection of different settings.

Site Review

When determining the allocation of land for specific purposes during the preparation, alteration or replacement of a development plan, the local planning authority should establish whether potential sites can be safely developed without the need for new or improved flood and coastal defences. (Figure 14.4). The checklist presented in Appendix A sets out a 3 stage procedure for this purpose which involves:

Stage 1; identification of vulnerable settings;

Stage 2; determining the flood and coastal defence needs associated with particular land uses or types of development. This should involve considering whether satisfactory flood or coastal defence infrastructure exists and can be maintained over the life of the development **or** that new defences can be provided by the developer that do not affect other interests.

Stage 3; local plan proposals should take due account of the nature and scale of any precautionary measures that may be acceptable. Prospective developers will thus be forewarned of the likely problems and can estimate the cost

Table 14.4 Summary of the preliminary planning guidance which may be appropriate for selected settings.

SETTINGS	DEVELOPMENT PLANS	DEVELOPMENT CONTROL
• actively unstable river channels • unprotected floodplains • rapidly eroding cliffs • actively unstable slopes • unprotected low lying areas • natural coastal defences (eg sand dunes) • very high – high sensitivity coasts	Areas least suited to development due to physical conditions. Local plan development proposals subject to major constraints.	Should development be considered it will need to be preceded by a detailed investigation, full risk assessment and/or environmental study. Many planning applications in these areas may have to be refused on the basis of potential physical problems.
• dry valleys in rural areas • eroding river channels • floodplains with low standard of defences • eroding cliffs • potentially unstable slopes • low lying areas with low standard of sea defences • sand dunes • saltmarsh areas • foreshores in important sediment transport zones • high – moderate sensitivity coasts	Areas likely to be subject to significant constraints due to physical conditions. Local plan development proposals should identify and take account of the nature of potential problems and address the requirements for suitable flood or coastal defences.	A site reconnaissance study will need to be followed by detailed site investigation, including risk assessment and/or environmental study prior to lodging a planning application.
• upland areas adjacent to stream channels and gulleys • areas behind eroding cliffs • estuaries • foreshores	Areas which may or may not be suitable for development but investigations and monitoring may be required before any local plan proposals are made.	Areas need to be investigated and monitored to determine risks, sediment budget, or sensitivity. Development should be avoided unless adequate evidence of suitable conditions is provided.

implications.

In many instances, this will involve no more than consultation with the appropriate local authority engineers or the NRA.

Investigation and Reporting

Where the local planning authority has reason to believe that some or all of a proposed development may cause or be affected by erosion, deposition or flooding, it should request the developer to submit a **Site Report**, either as a condition of granting outline consent or before determining the application. Site Reports should be prepared by a **competent person** who should be able to demonstrate relevant specialist experience in the assessment and evaluation of these processes. A competent person would normally be expected to be:

• a Chartered Geologist;
• a Member of the Institution of Civil Engineers;
• a Member of the Institution of Mining and Metallurgy; or
• a Member of the Institution of Water and Environmental Management.

The developer should be aware of the advantage of obtaining such reports from persons or organisations who possess and maintain professional indemnity insurance. This provision will promote a degree of commercial responsibility for the opinions expressed and will help to safeguard both the developer's and local planning authority's interests in the event of the development being damaged subsequently. Before making a Report, the competent person should undertake such inspections and investigations as are considered necessary to allow an opinion to be made on the safety of the development and its surrounding area.

In preparing the Site Report, the competent person should consider all factors which might influence the safety of the site and surrounding area in relation to its suitability for the proposed development. The report should therefore contain:

- a review of relevant background information;

- a factual record of investigations carried out;

- a description and engineering interpretation of relevant ground, surface water and groundwater conditions;

- an account of any stability or hydrological calculations;

- an account of any risk assessment, analysis of sediment budget or sensitivity;

- conclusions on the conditions of the site and the potential effects of the proposed development;

- recommendations for remedial, preventive and precautionary measures, site inspections and monitoring.

The local planning authority should examine the Site Report to check that its content and scope are in broad agreement with their requirements. It may be appropriate to seek assistance from other departments of the local authority in carrying out this check and it would be preferable if there was consistency in the review procedure. As it is the responsibility of the developer and his competent person to satisfy themselves that their development can be constructed without initiating problems, it should not normally be necessary for the local planning authority to check design assumptions and calculations. In the event of a dispute, however, the authority may have to seek advice from an outside agency.

It is important to recognise that most planning applications are for "**small developments**" (e.g. extensions to existing buildings and garages) or **changes in use** of a building. In many instances, these types of development need not be subject to a Site Report requirement provided that the site is not to be significantly disturbed and that the development would not interfere with the operation of erosion, deposition and flooding processes (see below).

The Effects of Development

Development may have a significant adverse effect on the level of risk from erosion, deposition and flooding at and around a site and elsewhere within a river catchment or coastal system. Local planning authorities can help ensure that these effects are minimised, most notably through requiring development proposals to take account of:

- **runoff control;** new development can significantly increase the quantity and rate of runoff reaching watercourses and, hence, may lead to increased flood risks in downstream areas.

  To minimise potential problems appropriate alleviation or mitigation measures should be included, such as the provision of surface water storage areas and flow limiting devises;

- **floodplain storage;** new development in floodplains can increase the risk of flooding elsewhere by reducing the floodwater storage capacity of the floodplains and/or impeding the flow of the waters. Appropriate mitigation measures should be taken to restore floodplain storage or provide artificial storage capacity;

- **coastal cliff management;** coastal cliffs can be very sensitive to the effects of land use and development with problems frequently associated with artificial recharge of groundwater levels through soakaways or leaking pipes, and inappropriate excavations. Much can be done to reduce the likelihood of slope problems, especially by the provision of positive drainage of surface water, avoiding the use of septic tanks and soakaways and using flexible service connections capable of sustaining small movements without leakage;

- **mineral extraction operations;** extraction of sand and gravel from the river bed, sand dunes, beaches and shingle structures may adversely affect the stability of these environments, increasing the rate of erosion or vulnerability to flooding. Mineral extraction in these areas will generally be inappropriate. Caution is also needed when restoring sand and gravel workings on river floodplains; the design of a restoration scheme for agriculture may lead to a reduction in floodplain storage if a raised, free-draining landform is required. In such circumstances, the local planning authority should discuss the proposals with the NRA (in England and Wales) or, in Scotland, seek the advice of the Regional Council Water Services Department (or equivalent).

Cooperation and Coordination

Erosion, deposition and flooding can give rise to strategic issues because the scale over which natural processes operate is extensive and may span regional and local authority boundaries. Indeed, the operation of a process in one area may have an impact elsewhere within a river catchment or coastal system. For example;

- hillslope erosion or riverbank erosion can lead to deposition within river channels downstream, reduced channel capacity and, hence, increased flood risk and navigation difficulties;

- coastal landslides can be an important sources of sediment for neighbouring beaches, sand dunes and saltmarshes. Disruption to the supply of sediment as a result of coast protection works may lead to a loss in value of conservation features and increased erosion or vulnerability to flooding, elsewhere.

Local planning authorities should work closely together on erosion, deposition and flooding issues, especially in preparing development plans within the same river catchment or coastal system to ensure that development plans contain consistent policies. Regional planning groups can provide a means of defining key issues and coordinating policies. It is especially important that local planning authorities liaise with those authorities and groups involved with the strategic planning of flood and coastal defences (i.e. the NRA, coastal defence groups and coast protection authorities) and conservation agencies; these bodies should be able to provide advice on technical matters of common interest. Local planning authorities should seek to ensure that development plan policies address land use issues of relevance to the flood and coastal defence strategy set out in the Catchment Management Plans, Water Level Management Plans and Shoreline Management Plans and are supportive of the conservation objectives for an area.

The Need for Monitoring

Global warming and sea level rise could have significant effects on the erosion, deposition and flood character of Great Britain, especially the possible increased frequency of extreme events. Changing patterns and occurrences of events may result in an expansion of the potentially vulnerable areas. Planning policies should, therefore, be kept under review to monitor their effectiveness in addressing erosion, deposition and flooding issues. **It is considered important that the risk maps used by local authorities are reviewed and updated to take account of changing conditions.** This should involve liaison with the NRA and other relevant authorities to:

- develop coordinated monitoring programmes, involving measurement of erosion and deposition rates;

- maintain detailed records of erosion, deposition and flood events.

Chapter 14: References

Buck A L 1993. An inventory of UK estuaries. 7 Volumes JNCC, Peterborough

Dargie T C D 1993. Sand dune vegetation survey of Great Britain, 3 Volumes. JNCC, Peterborough

Davidson N C, d'A Laffoley D, Doody J P, Way L S, Gordon J, Drake C M, Pienkowski M W, Mitchell R and Duff K L 1991. Nature Conservation and estuaries in Great Britain. Nature Conservancy Council, Peterborough

Doody J P, Johnston C and Smith B 1993. Directory of the North Sea Coastal Margin. JNCC, Peterborough

HR Wallingford, various dates. Macro Review of the Coastline of England and Wales. 10 Volumes.

HR Wallingford 1993. Coastal management: mapping of littoral cells. Report SR328.

MAFF, Welsh Office 1993. Strategy for flood and coastal defence in England and Wales. MAFF Publications.

Nature Conservancy Council, undated. Atlas of Coastal sites sensitive to oil pollution. English Nature, Peterborough

Rendel Geotechnics, in prep. Soft Cliffs: Prediction of Recession Rates and Erosion Control Techniques. Reports to MAFF.

Ritchie W and Mather AS 1985. The Beaches of Scotland. Countryside Commission for Scotland.

15 Summary and Conclusions

Introduction

The foregoing chapters have provided a broad outline of the nature and scope of hazard management in Great Britain, describing the legal and administrative framework (Chapter 3) the investigation techniques and sources of available information (Chapters 4 – 7) and the range of management options that are relevant in different environments (Chapters 8 – 12). The role of the planning system has been examined in Chapter 13, with a framework of advice for planners and developers presented in Chapter 14. The purpose of this Chapter is to summarise key aspects of investigation and management **in so far as they are relevant to the effective operation of the planning system in achieving the objectives of sustainable development.** This needs to be based on (See Chapter 4):

● the use of best possible earth science information;

● precautionary action to minimise risks;

● consideration of environmental impacts;

● the cost of investigation and precautionary measures should be borne directly by the developer.

A key theme throughout this Report has been how flooding and coastal erosion can present significant constraints to land use and development, imposing very high costs on society through damage to property, services and infrastructure, emergency relief and the resources devoted to management of the problems. Potential problems associated with other processes (e.g. hillslope erosion, channel instability and sedimentation in rivers and estuaries) should not be dismissed as trivial matters; they can impose significant costs on specific sectors of the economy (e.g. water

companies, navigation and harbour authorities) and increase the risk of potentially damaging flood events. The operation of these natural processes is, however, necessary for maintaining many landscapes, and nature conservation features in a sustainable condition. The dilemma for those involved in the management of erosion, deposition and flooding related issues is the need to reconcile the contrasting requirements of mitigating against their impacts and conserving the natural environment.

The broad scale operation of these natural processes dictate that it is generally inappropriate to address management issues on a site specific basis; strategic considerations are necessary on a catchment or coastal system–wide basis. Indeed, the complex operation of processes within these natural systems present a number of challenges to decision–makers:

● erosion, deposition and flooding problems are often linked and should not be treated in isolation. Flooding, for example, can initiate considerable erosion and deposition on hillslopes, rivers and on the coast. Flood problems are often exacerbated by erosion and deposition;

● development may lead to significant changes elsewhere within a system. These changes can affect the level of risk elsewhere or lead to the degradation of both natural landforms and the habitats which they support;

● the cumulative effects of development on a system may take many years to become apparent. Indeed, past responses to the threat of erosion and flooding have led to many long term problems facing river and coastal managers.

The administrative framework for erosion, deposition and flood management is a very complex system; it reflects the need to reduce the levels of risks faced by communities in vulnerable areas through flood warnings and defences, requirements to maintain effective and navigable waterways and obligations to take into account the interests of other groups such as conservation bodies and fisheries interests. It, therefore, extends beyond management of risks to include aspects of land and water resource management.

The framework may be visualised as comprising of 6 key elements:

(i)   common law rights to provide individuals with powers to protect their own property;

(ii)   enabling powers that give statutory authorities permissive or mandatory powers to undertake defence works for the benefit of the nation or maintenance operations;

(iii)   consenting arrangements which ensure that hazard management measures and land uses do not affect other interests or increase the level of risk elsewhere;

(iv)   provisions to ensure the conservation and enhancement of landscape and nature conservation features, involving the protection of designated sites and areas of national and international importance, and the setting of statutory environmental duties on various operating authorities and harbour authorities;

(v)   the establishment of specific national agencies to exercise supervision over all matters related to particular processes;

(vi)   consultation arrangements between key interest groups whose interests may be affected by hazard management measures.

The principal characteristics of the management frameworks in England and Wales, are significantly different than those in Scotland, most notably with respect to the arrangements for flood defence (there is no equivalent to the NRA in Scotland). The evolution of these systems does, of course, reflect the nature and scale of particular problems in the different countries within Great Britain. In much of Scotland, for example, the level of flood risk is generally lower than in parts of England and Wales where extensive areas of development are often concentrated on floodplains. However, the response

to the 1990 and 1993 floods in Tayside (see Rendel Geotechnics, 1995a) suggest that there may be increasing pressure for a more sophisticated flood management system in Scotland; possible future changes could address the need for a comprehensive approach to river channel and runoff management (see Chapters 10 and 11) and the establishment of an agency to supervise flood defence considerations on a catchment-wide basis.

Bearing in mind the potential effects of hillslope management practices on water quality, flood risk, river channel behaviour and river corridor habitats there are arguments for suggesting that consenting arrangements may ultimately need to be established to regulate hillslope erosion control issues. This is particularly so in England and Wales, where the NRA's responsibilities for the water environment includes water quality and resources, extending beyond watercourses to address problems such as nitrate pollution and groundwater protection. Indeed, effective catchment management may be constrained by the present limited influence over activities away from watercourses.

The planning system can have a central role in the management of erosion, deposition and flooding issues, most notably through ensuring that:

● development is not placed in unsuitable locations;

● development does not increase the degree of risk elsewhere;

● conservation interests are taken into account in the selection of hazard reduction measures;

● management plans are supported by appropriate development plan policies.

However, there is a difference between what can be achieved and what is achieved by many local planning authorities, with around 50% of surveyed authorities having no policies addressing any aspects of risk management issues (Chapter 13). The reasons why the planning system has not realised its potential are likely to be complex, but reflect the way the system operates rather than deficiencies in the legislative provisions. Amongst the more important factors is the limited nature of existing planning guidance which does not clearly set out the full range of ways in which the system can be used or address the relationships between local planning authorities and flood and coastal defence interests. In Scotland, the absence of specific guidance addressing these issues may be a major factor in limiting the effectiveness of the system.

Earth Science Information Needs

Planners and developers need to have access to a range of technical information about erosion, deposition and flooding processes if the potential conflicts between hazard management and conservation are to be adequately reconciled. However, the nature of the investigation techniques and the level of detail will vary with the different scales of decision making. For example planners need to address these issues at a regional and strategic level, in the preparation of development plans and the determination of planning applications.

The planner's requirement is for **general guidance** on where hazard events are likely to occur, the nature of the potential problems and the possible effects of the development and any precautionary works on the natural environment. Planners also need to know, in general terms, the types of development that might be preferred or opposed in particular areas. In addition, they need to have sufficient appreciation of potential problems in an area to decide whether adequate site investigation information has been submitted with a planning application. This need is defined by the Town and Country Planning Act 1990 (and the 1972 Act in Scotland), with local planning authorities **required** to keep under review the principal physical characteristics of their area and, where necessary, neighbouring areas.

The primary responsibility for determining whether a site is suitable for a particular purpose rests with the developer who needs to:

● decide, in broad terms, where development might be most suitably and economically located;

● assess the feasibility of a project at short listed or chosen sites;

● satisfy the planning authority that the development proposal has taken account of potential problems associated with risks and the effects of defence works on the environment.

The developer's main requirement is for detailed information about the ground conditions at particular sites, the risks that can be expected over the lifetime of the development and whether any problems can be economically overcome. However, developers also need general information for preparing desk studies in advance of site investigations and to be able to take account of the potential effects of the proposed development elsewhere within a catchment or coastal system.

In many ways, the information requirements of the various bodies and organisations involved in managing aspects of the effects of erosion, deposition and flooding are similar to those of developers. Detailed information is required for individual sites to determine where resources should be best directed and the most appropriate management option. However, it is now widely appreciated that decision-making about individual sites needs to take place within the context of broad strategic plans prepared for individual catchments, wetlands or coastal systems. Preparation of management plans needs to be supported by a general awareness of the operation of physical processes within these systems, together with an appreciation of the issues that may need to be addressed in particular areas.

For all those involved in the management of erosion, deposition and flooding issues, the requirement for technical information flows from a general awareness of the potential constraints and resources in an area to the need for site specific information. The general model of investigation presented in Chapters 4–7 involves three contrasting but complementary levels of investigation:

(i) **general assessment** of broad areas at national, regional and local levels;

(ii) **site reconnaissance** of selected sites or specific vulnerable areas at local and site level;

(iii) **detailed investigations** at site level.

A number of important principles are relevant at each scale or stage:

● investigations should consider both the "**site**" (i.e. the area of interest) and the "**situation**" (i.e. the area over which active processes can contribute to potential problems at the site **or** the area over which the effects of development can be transmitted). This requires an understanding of physical systems which can operate at various hierarchical scales from, for example, littoral cells on the coast down to individual landforms;

- investigations should consider the historical record of past events within the area of interest and beyond, to gain an appreciation of the nature and scale of the problems that can arise, their impact, and to obtain an indication of how frequently such events could be expected (i.e the **risks**);

- investigations should be taken account of the importance of sediment supply and transport in determining the sustainability of particular river corridor or coastal landforms (i.e the **sediment budget**);

- investigations should consider the potential for change within the area of interest and the **sensitivity** of individual features to the effects of climate change (e.g. sea level rise) or the effects of development.

Investigations may be needed to **monitor** the erosion, deposition and flooding character of an area; to keep strategic and detailed policies under review, to assess the effects of development on the environment, to provide essential information on the nature of problems at a site or to evaluate the effectiveness of particular management strategies. Attention is drawn to the fact that, for coastal areas in England and Wales, future monitoring requirements will be addressed in Shoreline Management Plans (MAFF, 1994).

A wide variety of techniques are available for measuring erosion, deposition and flooding processes and the individual factors which may control their occurrence in different environments. For example, a combination of weather radar, rain gauges, flow gauges and marker boards may be used to assess flood risk on rivers and provide early warnings of potentially damaging events. Cliff monitoring is widely practised to provide a means of accurately and objectively gauging the nature and rate of cliff recession. Foreshore conditions can also be monitored to provide an indication of the changing vulnerability of a cliff to wave attack as beach levels rise or fall.

The selection of measurement techniques will be influenced by the scale of decision-making and the nature of the environment. It is, therefore, necessary for managers to develop monitoring

strategies at a variety of scales. The choice of techniques or combination of techniques will, however, be governed by the objectives of the measurement exercise. In many circumstances, for example, the recording of the cumulative loss of land on an annual basis may be appropriate, using remote techniques (e.g. aerial photographs) or simple ground survey methods. Elsewhere detailed recession measurement may be necessary to predict future recession rates at particular sites and monitor the effectiveness of particular management strategies.

The local planning authority is entitled to required the developer, at his expense to provide suitable expert advice and relevant information with a planning application, and is entitled to rely on that advice in determining the application and formulating any necessary conditions. However, it is important that development plan preparation is supported by an appropriate level of information on the nature of physical conditions in an area. Failure to recognise potential problems at an early stage of the decision making process can lead to increased development costs, abandoned development, calls for publicly funded defence works, damage to adjacent land, property and the environment or an increase in the risks to other land users. Policy omissions can result in development proceeding in areas which might not be suitable.

A 5-stage approach to undertaking a general assessment of the physical characteristics of an area has been set out in Chapter 14 which involves:

- consideration of key issues;
- identification of key information sources;
- assessment of the degree of risk;
- assessment of the potential environmental effects;
- presentation of information.

Table 14.3 provides a checklist of some of the key sources of information, most of which are directly available from the relevant organisation. However, a number of points are worthy of note regarding their reliability:

- a number of sources are reviews compiled from primary information that is likely to be of variable quality (e.g. the National Landslide Databank; the Macro-Review of the Coastline; the JNCC Directory of the North Sea Coastal Margin). In some instances the information contained within these sources may have to be verified or extended to provide a reliable statement of the physical characteristics of an area;

- many map sources have been compiled for purposes other than land use planning. The significance of the various map units to planning may be unclear and careful professional judgement is likely to be needed to **derive** relevant information from these sources. For example, geological units containing bulk materials such as sand and gravel can be inferred from available geological maps and memoirs.

Management Plans and NRA Surveys of flood risk areas are currently in preparation or scheduled for a future date. They must, be regarded as important **potential** sources; it is not possible, however, to evaluate their availability or reliability until such time as they are completed and become available for use by decision makers.

Local planning authorities should make full use of the expertise available in other departments within the local authority (e.g. coastal engineers and water engineers) and the NRA. For specialist subject areas it may be necessary to seek advice from commercial consultants.

The accessible presentation of information is an important part of any general assessment. However, few planners have a background in the earth science and the technical nature of traditional geological and geomorphological results have led to these sources being under–used and under–appreciated. There is also a need to consider the issues associated with erosion, deposition and flooding within the broader context of the constraints and resources imposed by the ground conditions in an area, especially the potential for housing and employment opportunities.

In recognition of the difficulties in incorporating earth science information into the planning process has led the DoE to actively promote the development of Applied Earth Science Mapping as a vehicle for presenting technical information to the planning community. The techniques involve collecting and collating available earth science information from various sources and then summarising it in a form specifically tailored to meet local planning needs. In general, this type of study is directed towards preparing a combination of **thematic maps** at a general scale (1:50,000 – 1:10,000 scale), which become increasingly focused on key planning issues as they are developed from the basic factual information. A wide range of

maps can be custom made on specific topics and for specific audiences. By and large planners are served by summary and derived maps which allow them to devise broad, strategic policies for development and conservation and indicate, in general terms, areas in which certain types of development might be preferred or opposed (See Chapter 14).

Measures for Risk Reduction

A range of responses are available for the management of erosion, deposition and flooding hazards, including:

- **Acceptance** of the risk;

- **Reducing** the occurrence of potentially damaging events;

- **Avoiding** vulnerable areas;

- **Protecting** against potentially damaging events.

In most cases, however, the response will be complex, involving a variety of measures adopted by different organisations at different locations within a catchment or coastal system. The most appropriate option will depend on the nature of the problem, the level of acceptable risk, the availability of resources and the statutory powers available to interested bodies or authorities to tackle the problems.

Hazard reduction cannot be separated from other aspects of environment management. Indeed, the Government environmental strategy (DoE, 1990) identifies three critical issues which are important influences on the selection of the most appropriate response:

- the need to resolve the conflicts between pressures for development and mobility, and the conservation of what is best in the environment;

- the need to maintain economic growth without making excessive demands on natural resources;

- the need to combat the dangers of pollution without jeopardising economic growth.

The potential conflicts that can be generated by the selection of hazard reduction responses (Table 15.1) emphasises the need for an approach to decision-making, which involves consultation and partnership between the relevant interests. Of particular importance is the need for coordination between land use planning, conservation, and flood and coastal defence. Management plans, (for catchments, water levels in wetlands and the shoreline) have the potential for providing a mechanism through which acceptable solutions can be identified and put into practice.

The degree of **risk** involved with development in vulnerable areas should be carefully considered by decision makers. Planning and management strategies should be developed that relate to these risks. A central requirement of most management approaches to erosion, deposition and flooding problems is to identify where damaging events of a particular size are likely to occur, the possible consequences and how often they can be expected.

Vulnerable areas can be readily identified from a range of sources that are currently available or are in preparation, (See Table 4.1). However, there are major problems in evaluating the potential for hazardous events to occur. Damaging events are often associated with very large return period storms (the Lynmouth floods of 1952 had an estimated return period of thousands of years) and are generally the product of a combination of circumstances (e.g. antecedent rainfall conditions, groundwater levels, land management practice).

Two contrasting approaches are used to predict events: **deterministic** and **probabilistic** (see Chapter 4). However, the accuracy of both approaches can be constrained by limited data sets from which assessments have to be made. Benson (1960) demonstrated that to achieve 95% reliability on the estimate of discharge for a 50 year flood event required 110 years of records. Such lengthy data sets are not common in Great Britain; detailed rainfall records are usually available for 100–150 years, hydrological records from river gauging stations generally for 20–40 years. One of the longest river gauging records cited in the Flood Studies Report (NERC, 1975) is for the Thames at Teddington, beginning in 1883.

Return period statistics can be unreliable; especially when derived from very short periods of records as new events can lead to a significant modification in the calculated return period. For example, the Truro floods of January 1988 initially had an estimated return of 350 years, using the procedures

recommended in the Flood Studies Report (NERC, 1975); a reappraisal of the risk after further floods in October 1988 suggested a return period of 50 years for the January event (Acreman, 1989).

The general public frequently misinterpret return period statistics, especially the chance of rare events occurring in a particular year. For this reason the degree of risk associated with flood events is now also expressed in terms of the **standard of protection** which exists (see, for example, MAFF 1993, Table 11.7). In many instances, the standard of protection can give the general public a clearer indication of the relative risks in different areas as potentially vulnerable areas can be defined in terms of the land use band which has the closest indicative standard of protection.

Notable difficulties also arise in the prediction of erosion rates, especially on coastal cliffs. Cliff recession is a episodic process; there may be little or no erosion for a long period, followed by sudden rapid recession in response to a large storm. The "rate" of recession may, therefore, depend on the timescale chosen. Furthermore, the main causal factors such as wave undercutting are inherently unpredictable. This uncertainty creates problems for coastal managers when attempting to define the degree of risk to cliff top property, either for identification of "set-back" lines by planners or the choice of erosion control technique by engineers. In this context, it is worth noting that MAFF have recently established a programme of research to develop methods of predicting soft cliff recession rates suitable for both detailed design and strategic planning (Rendel Geotechnics, 1995c).

The nature of the erosion, deposition and flood hazard can also be defined in terms of the potential for damaging events or the behaviour of the relevant physical system. This assessment should be based on a thorough review of available records, documents and reports, together with appropriate field investigations. Such studies can form the basis for planning guidance, as was the case for the investigation of coastal landslide problems at Ventnor, Isle of Wight (Figure 6.5; Table 6.6). In this instance, the planning and management framework was based on an understanding of the nature, magnitude and frequency of contemporary ground movement processes and their impact on the local community (Lee and Moore, 1991). There is great potential for adapting this type of study to flood prone areas, to demonstrate how flood risk information can be incorporated into the into the planning process.

Table 15.1 A summary of some of the main resource management issues associated with flood and coastal defence.

ACTIVITY	ISSUES OF CONCERN
Sea Users	• concerns that coastal defences may restrict access for recreational sea users.
Use of Navigable Waterways	• concerns over the presence of potentially contaminated sediments in dredged material may constrain channel maintenance operations. • the effects of dredging on flood and coastal defence interests may lead to restrictions in operations. • flood defences may lead to conflicts with recreational water users and fisheries interests.
Conservation	• defences may cause visual intrusion and lead to degradation of nature and geological conservation interests. • defences may result in disturbance of marine or river corridor archaeological sites.
Minerals	• the potential for erosion and increased flood risk may constrain aggregate extraction from river channels, sand dunes, beaches and the sea bed. This places additional pressure on alternative sources. • the need to maintain floodplain storage may restrict opportunities for the restoration of mineral workings to agriculture.
Agriculture and Forestry	• concerns that land management practices can have an effect on flood risk elsewhere.

The problems in accurately defining the level of hazard in an area, together with limited knowledge of the costs incurred by past events make conventional **quantitative risk assessment** techniques (Chapter 4) difficult to adapt to erosion, deposition and flooding problems. However, the **relative risk** approach outlined in Chapter 6 may have wider use in comparing the scale of the problems at a number of potential sites; such sites could be under consideration for development plan proposals or on a developers shortlist of potential investment sites.

Assessment of Environmental Effects

It is important that the potential environmental effects of development are considered at all stages of the decision making process. These effects can be complex, involving modifications to the level of risk, changes to the natural environment or the land use character of an area (see Figure 15.1). At a strategic level decision makers should be aware of the distribution of conservation and recreation resources in an area, and the extent to which their **sustainability** is related to the continued operation of physical processes.

At a site level, the assessment of environmental effects (either as part of a formal EA or to provide necessary supporting information with a

development proposal) should involve the identification of:

• the potential effects of the proposed development on the operation of physical processes;

• those landforms, habitats or features that are sensitive to changes in the rate of physical processes;

• the possible consequences of these changes on the natural environment;

• measures that can be employed to minimise the adverse effects on the environment.

In many instances it may be of benefit to undertake a preliminary environmental study (a **scoping study**) to establish the key concerns that are likely to arise **before** the allocation of land, by planners, for development in vulnerable areas or the selection of sites for investment, by developers. This type of study should involve consultation with key interest groups and authorities and the qualitative appraisal of possible impacts (see the impact matrix presented in Figure 6.6).

Figure 15.1 Risk Management: a summary of key issues in the coastal zone (after Rendel Geotechnics, 1993).

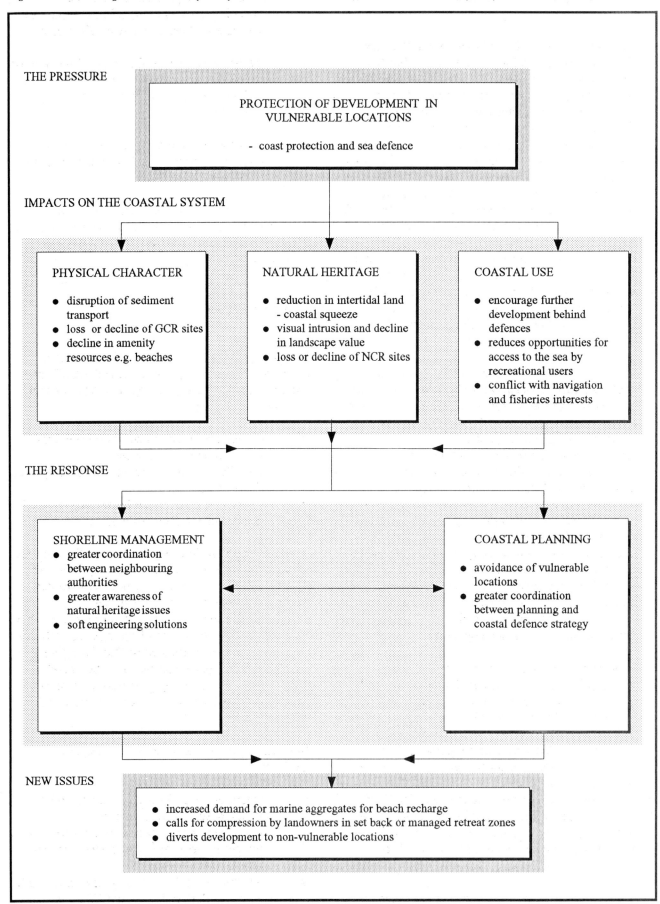

THE PRESSURE

PROTECTION OF DEVELOPMENT IN
VULNERABLE LOCATIONS

- coast protection and sea defence

IMPACTS ON THE COASTAL SYSTEM

PHYSICAL CHARACTER

- disruption of sediment
  transport
- loss or decline of GCR sites
- decline in amenity
  resources e.g. beaches

NATURAL HERITAGE

- reduction in intertidal land
  - coastal squeeze
- visual intrusion and decline
  in landscape value
- loss or decline of NCR sites

COASTAL USE

- encourage further
  development behind
  defences
- reduces opportunities for
  access to the sea by
  recreational users
- conflict with navigation
  and fisheries interests

THE RESPONSE

SHORELINE MANAGEMENT
- greater coordination
  between neighbouring
  authorities
- greater awareness of
  natural heritage issues
- soft engineering solutions

COASTAL PLANNING

- avoidance of vulnerable
  locations
- greater coordination
  between planning and
  coastal defence strategy

NEW ISSUES

- increased demand for marine aggregates for beach recharge
- calls for compression by landowners in set back or managed retreat zones
- diverts development to non-vulnerable locations

The Developer Pays Principle

The unwillingness of local planning authorities to view erosion or flooding as anything other than a problem for the developer, has led to the necessity of managing catchments or coastlines through engineering options rather than planning control. Paradoxically, it may have been the clear success of engineered defences that has led to a lack of control of development in those areas at risk from natural hazards. For example, construction of flood defences often leads to increased pressure for development in what is now perceived to be a safe area. In reality, however, construction of flood defences only **reduces** the risk of damage. **It cannot eliminate the risk.** Increased investment and density of development behind the defences may only lead to higher losses when, inevitably, larger floods occur and to more pressure for expensive maintenance and replacement of defences as they become worn.

Owners of property in vulnerable areas naturally apply pressure on local authorities to provide flood or coastal defences to safeguard their investments. Such defences are mainly funded by the public purse through grant aid and local authority contributions. The defences can, however, be the source of conflict with environmental interests. It is now appreciated that to attempt to provide universal protection would be uneconomic and work against the operation of physical processes, creating problems elsewhere. Government policy, therefore, is to avoid placing an additional burden of responsibility on future generations by increasing unnecessarily the number of areas to be artificially protected. (MAFF, 1993). For this reason, grant aid for flood and coastal defence is not available for new development.

Bearing in mind that there is essentially a presumption in favour of development proposals that are in accordance with the development plan it is important that plan policies and proposals take full account of flood and coastal defence issues. In seeking to decide where development might be most suitably located, local planning authorities will need to establish whether a shortlisted site can be safely developed without the need for flood defence improvements. Should new defences or improvements be required to protect land allocated for development, the authority may need to consider, in general terms (Figure 15.2):

- whether the necessary new defences or improvements can be provided without

adversely affecting other interests, including conservation, mineral resources and coastal defence interests elsewhere;

- the possible cost implications of the new or improved defences to potential developers and whether they are likely to be economical;

- the possible nature and scale of new or improved defences that would, in principle, be acceptable.

It will, however, be the responsibility of the developer to fully investigate a site and design and build any necessary defences (after receiving appropriate approvals).

Recommendations

In this Report it has been shown how erosion, deposition and flooding can represent significant problems for land use planning and development, most notably through:

- the impact of the processes on property, infrastructure and services;
- the effects of development on the degree of risk elsewhere;
- the conflicts generated by the selection of hazard management strategies.

In areas where erosion, deposition and flooding are likely to impose constraints to development and land use, decision makers will need to consider identifying those areas where particular consideration should be given to these issues. They may also need to be aware of the type of problems that may occur, how frequently that damaging events may take place and whether they can be satisfactorily overcome by proposed developments or uses without leading to adverse environmental effects.

In general, planners and developers require guidance on:

- where erosion, deposition and flooding issues need to be considered;
- the nature of the hazards in the areas and the general risk to existing land uses and development;
- whether potentially damaging events can be expected during the normal lifetime of particular types of development;

191

Figure 15.2 Identification of opportunities for development in river corridors and the coastal zone, taking into account physical processes.

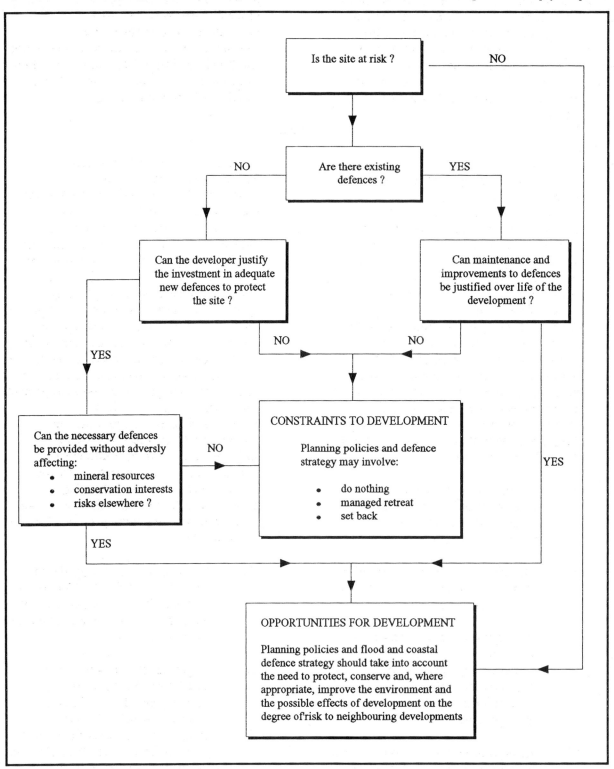

- the types of uses or development that, in general terms, may be best suited to those areas;
- whether development or land use can have adverse effects on other interests in the river or coastal environments.

As described earlier this requires access to reliable technical information and advice from authorities and bodies with an interest or responsibility in various aspects of environmental management. However, an equally pressing need is for local planning authorities to be aware of the ways in which the planning system can be used to achieve management objectives. In the past, this potential has been seldom fully realised. There is, therefore, a clear need to improve the awareness of the role of the planning system as an instrument in the management of erosion, deposition and flooding issues. This should involve considering the following recommendations (in conjunction with those recommendations presented in Rendel Geotechnics (1995b):

Legislation and its Use

1. The existing planning legislation sets out a flexible framework to regulate the development and use of land in the public interest. Although many of the powers available to local planning authorities are essentially restrictive in nature, authorities also have the opportunity to act in a positive manner through the use of discontinuance orders or compulsory purchase. These powers appear not to be used to address erosion, deposition and flooding issues. **It is recommended, therefore, that the government considers undertaking a review of these powers and evaluates their potential as an effective tool in managing risk problems along river floodplains and in the coastal zone.** This could involve:

   1a   Establishing the extent to which these powers have been used by local planning authorities, and in what circumstances;

   1b   Identifying the possible limitations to using these powers as a tool in risk management;

   1c   Identifying the circumstances in which the use of such powers would be appropriate and providing a general indication of the geographical significance of such an approach;

   1d   Reviewing how the powers can be used to support coastal defence strategies set out in Shoreline Management Plans.

2. The **Building Regulations** are made to secure the health, safety, welfare and convenience of persons in and about the building. They provide a complementary mechanism to the planning system for addressing hazards during development. At present, however, the Regulations do not cover flooding problems. **It is recommended, therefore, that consideration is given towards extending the Building Regulations to include the need to take account of potential flood problems in building design.**

Planning Guidance

3   The current guidance to planners in England and Wales concentrates on addressing the most significant problems (flooding and coastal erosion) and establishes the important principle of avoiding areas at risk. However, there are a number of important omissions from the guidance which are seen to limit the current effectiveness of the planning system. The most notable omissions are the way the planning system considers the interactions between processes throughout catchments and coastal systems, and the complex range of direct and indirect impacts that the processes and management responses can generate. **It is recommended, therefore, that the Government considers preparing guidance for planning authorities in England and Wales which addresses erosion, deposition and flooding processes as important elements of the landscape and not merely risks.** This could involve:

   3a   Revising the existing guidance presented in PPG 14, PPG 20 and Circular 30/92, taking account of the framework of advice set out in Chapter 13;

   3b   Preparing specific guidance on **Catchment Planning** (to complement PPG 20 Coastal Planning) which addresses catchment–wide issues and the need for liaison with the NRA over the preparation and use of Catchment Management Plans.

4   In Scotland, the absence of planning guidance on matters relating to erosion, deposition and flooding is seen as a significant constraint on the effective operation of the planning system. **It is recommended, therefore, that the Scottish Office considers preparing**

guidance which addresses the physical risks to development. This could include:

4a   Preparing guidance on flood risk matters (NB. Draft guidance was issued for consultation in March 1995);

4b   Revising existing Coastal Planning Guidelines to take account of areas vulnerable to flooding and coastal erosion, and the potential effects of sea level rise.

Sources of Information, Cooperation and Coordination

5   It is recognised that conservation agencies, the NRA and coastal defence groups can have an important role in providing technical advice to local planning authorities in England and Wales. Management plans for catchments, the coast and the shoreline are likely to be an effective mechanism for coordinating the various interests relating to erosion, deposition and flooding issues. The preparation of these plans is, however, in its infancy. **It is recommended, therefore that the Government undertakes a review of the effectiveness of these plans once the programme of plan preparation is complete and experience has been gained of their use as management tools.** This could involve reviewing:

5a   the extent to which management plans have taken account of land use issues;

5b   the effectiveness of management plans as a source of technical information for planners;

5c   the extent to which the preparation and use of management plans has helped resolve conflicts between conservation, land use, and flood and coastal defence interests;

5d   the extent to which management plans have been supported by the preparation of development plan policies and proposals.

6   The ad hoc arrangements for catchment and shoreline management in Scotland can be viewed as a limitation to effective strategic planning of flood and coastal defence matters, and is a constraint to the consideration of related land use planning issues at a strategic level. **It is recommended, therefore, that the Scottish Office considers encouraging and supporting a programme of management plan preparation, led by Regional Councils and the River Purification Boards.**

7   The preparation of S.105 flood risk maps by the NRA in England and Wales is likely to make an important contribution to improving the effectiveness of the planning system. However, in many areas the S.105 surveys have not been undertaken or are in progress. There will be a need, therefore, to monitor the effectiveness of this approach as a means of increasing awareness of flood risk amongst the planning community. **It is recommended, therefore, that the Government undertakes a future review of the effectiveness of the S.105 flood risk maps.** This could involve reviewing:

7a   the extent to which the maps could be used to highlight those areas that may be sensitive to the effects of flood defence works;

7b   the extent to which the flood risk maps have identified areas where rare events could lead to severe losses i.e. downstream of dams, in confined upland valleys etc;

7c   the extent to which the maps could be used to highlight those areas that may be sensitive to the effects of flood defence works.

8   The absence of readily available flood risk information in Scotland is an important constraint to addressing flood issues in the planning process. **It is recommended, therefore, that a programme of systematic surveys of areas where flood problems occur is undertaken in urban and rural areas of Scotland.**

9   Despite the success of Applied Earth Science Mapping in areas where the DoE have commissioned such studies, many local planning authorities appear unaware

of the techniques as an effective means of information collections and presentation. **It recommended, therefore, that the DoE prepares guidance to local planning authorities on the use of Applied Earth Science Mapping techniques.** This could involve:

9a   Preparing a handbook for undertaking Applied Earth Science Mapping;

9b   Drawing the techniques to the attention of local planning authorities when relevant planning guidance is next revised.

10   The study of Coastal Landslip Potential Assessment, Ventnor Isle of Wight outlined range of approaches for responding to coastal landslide problems which provide a basis for planning and development decisions in the town. It is recognised that the study has broader applicability in providing a general approach to presenting hazard potential and risk information in suitable formats for use by planners and developers. **It is recommended, therefore, that the Department considers establishing similar demonstration projects which address other potential constraints to development, including:**

10a   the occurrence of extremely rare but potentially very damaging events such as flash floods or mudfloods in dry valleys;

10b   the potential for lowland floods or coastal floods.

11.   Sediment budgets and sensitivity are recognised to be important considerations for sustainable planning and management on a catchment or coastal system basis. **It is recommended, therefore, that the Department commissions a review of the best ways that these specific considerations can be addressed by local planning authorities and developers.**

Chapter 15: References

Acreman MC 1990. Flood Frequency analysis for the 1988 Truro Floods. Journal of the Institution of Water and Environmental Management 4, 62–69

Benson MA 1960. Characteristics of frequency curves based on a theoretical 1,000 year record. In Flood Frequency Analysis, USGS Water Supply Paper 1543 –A.

Department of the Environment 1990. This Common Inheritance. Britain's Environmental Strategy. HMSO.

Lee EM and Moore R 1991. Coastal landslip potential assessment, Ventnor, Isle of Wight. Report to DoE.

MAFF 1993. Flood and Coastal Defence: Project Appraisal Guidance Note. MAFF Publications.

MAFF 1994. Shoreline Management Plans: A Guide for Operating Authorities (consultation Draft, October)

NERC 1975. Flood Studies Report. Institute of Hydrology, Wallingford.

Rendel Geotechnics 1993. Coastal Planning and Management: a review. HMSO.

Rendel Geotechnics 1995a. Erosion, Deposition and Flooding in Great Britain Methodology Report. Open File Report held at the DoE.

Rendel Geotechnics 1995b. The Occurrence and Significance of Erosion, Deposition and Flooding in Great Britain Report to the DoE.

Rendel Geotechnics 1995c. Soft Cliffs: Prediction of Recession Rates and Erosion. Control Techniques. Project Definition Report. MAFF

Scottish Office Environment Department 1995. Planning and Flooding. National Planning Policy Guidance (draft).

Appendix A: Checklists for Consideration of Erosion, Deposition and Flooding Issues by Planners

Guidelines for Earth Science Considerations in the Preparation of Development Plans

These guidelines have been prepared to assist those involved in carrying out surveys of an area in support of the preparation of development plans.

1.   **Identification of Vulnerable Settings and Key Issues**

In considering the occurrence of vulnerable areas and key issues have the following bodies should be consulted.

Flooding	Estuary Erosion and Deposition	Coastal Erosion
• NRA	• NRA	• Coast Protection Authority
• Internal Drainage Board	• Internal Drainage Board	• Coastal Defence Group
• Local Authority Coastal Engineers		

(a)   These issues are commonly associated with particular settings, you should consult with the relevant operating authorities and conservation agencies to establish whether they are significant in the area.

TERRAIN UNIT	RISKS	SEDIMENT BUDGET	SENSITIVITY
Mountainous and Upland Areas	• upland soil erosion • flash floods • sedimentation	• loss of water storage capacity in reservoirs • sedimentation in watercourses, increased flood risk and navigation problems	• land degradation and loss of amenity • creation of conservation sites in flood events
Undulating Lowlands	• soil erosion (water & wind) • flash floods and mudfloods • lowland floods • unstable river channels • river erosion and deposition	• sedimentation in watercourses, increased flood risk and navigation problems • disposal of dredgings on land • effect of mineral extraction from channels on erosion and flood risk • effect of mineral stockpiles on floodplain storage	• creation and maintenance of conservation sites • effects of channel maintenance and flood defence on conservation sites
Estuaries	• flooding • channel erosion • sedimentation	• sedimentation in estuary channels may lead to increased navigation problems • disposal of dredgings on land or at sea • sedimentation important to maintain mudflats and saltmarshes as natural flood defences	• flooding and deposition important to maintain and create conservation sites • sea level rise may lead to increased flood risk • channel maintenance and flood defences may affect conservation sites
Coastal Lowlands	• flooding	• sediment supply needed to maintain beaches, sand dunes, mudflats and saltmarshes as natural coastal defences • sediment supply needed to maintain amenity beaches and sand dunes • mineral extraction from beaches and dunes may lead to increased flood risk • sediment may be needed for beach recharge schemes	• sediment supply needed to maintain and create conservation sites • flood defences may affect conservation interests • sea level rise may lead to increased flood risk
Coastal Clifs	• landsliding • cliff recession	• supply of sediment from eroding cliffs needed to maintain coastal defences elsewhere • supply of sediment from eroding cliffs needed to maintain amenity beaches and sand dunes elsewhere • effects of mineral extraction on coastal slope stability	• erosion needed to maintain and create nature and geological conservation sites • coast protection may affect conservation interests • sea level rise may lead to accelerated erosion rates
Sand Dunes	• wind erosion • flooding	• sediment supply needed to maintain sand dunes as natural coastal defences • sediment supply needed to maintain amenity sand dunes • effects of mineral extraction on erosion and flood risk	• sediment supply needed to maintain conservation sites • wind erosion needed to sustain ecological value • coast protection may affect conservation interests • mineral extraction may affect conservation interests • sea level rise may lead to accelerated erosion

(b)   Do potentially vulnerable settings occur in an area?

TERRAIN UNIT	Y/N	KEY PROCESSES AND VULNERABLE SETTINGS (tick)		IS THIS A SIGNIFICANT PLANNING CONSTRAINT?
Mountainous and Upland Areas		Upland Soil Erosion	● Upland peaty soils ● Areas of recreation pressure ● Recently afforested land	
		Flash Floods	● Land adjacent to streams ● Areas downstream of dams and reservoirs	
		Sedimentation	● Upland lakes and reservoirs	
Undulating Lowlands		Water Erosion	● Areas of poor land management associated with silty and fine sandy soils	
		Wind Erosion	● Areas of poor land management associated with fine sandy soils	
		Flash Floods	● Land adjacent to tributary streams, especially around the margins of upland areas. ● Areas downstream of dams and reservoirs	
		Unstable River Channels	● Alluvial rivers around the margins of upland areas ● Alluvial rivers upstream and downstream of channelisation, river engineering works, bridges etc.	
		Mudfloods	● Dry valleys in areas of poor land management	
		Lowland Floods	● Low lying land adjacent to rivers and streams	
		Sedimentation	● Within alluvial river channels	
Estuaries		Flooding	● low lying land	
		Channel Erosion	● alluvial river channels	
		Sedimentation	● mudflats and saltmarshes ● within channel	
Coastal Lowlands		Flooding	● low lying land, including land behind sand dunes shingle ridges etc.	
Coastal Cliffs		Landsliding and Cliff Recession	● soft rock cliffs	
Sand Dunes		Wind Erosion	● unstabilised bare sand areas	

2.   Identification of Information Sources

The following are likely to be the key sources of information in particular terrain units. Please tick if they have been accessed:

MOUNTAINOUS AND UPLAND AREAS

- BGS geology maps and memoirs
- Soil Survey and Land Use Research Centre soil maps and memoirs for England and Wales
- Macaulay Land Use Research Institute soil maps and memoirs for Scotland
- NRA surveys of flood risk areas
- NRA Catchment Management Plans
- Conservation agency maps and records

UNDULATING LOWLANDS

- Soil Survey and Land Use Research Centre soil maps and memoirs for England and Wales
- Macaulay Land Use Research Institute soil maps and memoirs for Scotland
- NRA surveys of flood risk areas
- NRA Catchment Management Plans
- British Geological Survey maps, reports and records
- Conservation agency maps and records

ESTUARIES

- NRA flood risk maps
- NRA Sea Defence Survey
- Catchment Management Plans
- Estuary Management Plans
- Water Level Management Plans
- Shoreline Management Plans
- BGS geology maps and memoirs
- JNCC Inventory of UK Estuaries
- Macro Review of the Coastline (HR Wallingford)
- JNCC Directly of the North Sea Coastal Margin
- Conservation agency records and maps

COASTAL LOWLANDS

- NRA flood risk maps
- NRA Sea Defence Survey
- Shoreline Management Plans
- Catchment Management Plans
- BGS geology maps and memoirs
- Macro Review of the Coastline (HR Wallingford)
- JNCC Directory of the North Sea Coastal Margin
- Conservation agency records and maps

COASTAL CLIFFS

- National Landslide Databank
- Coast Protection Surveys
- Shoreline Management Plans
- BGS geology maps and memoirs
- Macro Review of the Coastline (HR Wallingford)
- JNCC Directory of the North Sea Coastal Margin
- Conservation agency records and maps
- Aerial photographs
- Historical topographic maps

SAND DUNES

- NRA flood risk maps
- Shoreline Management Plans
- BGS geology maps and memoirs
- Soil Survey maps and memoirs
- Macro Review of the Coastline (HR Wallingford)
- JNCC Directory of the North Sea Coastal Margin
- Conservation agency records and maps
- JNCC Sand Dune Vegetation Survey
- Meteorological data

EROSION, DEPOSITION AND FLOODING GENERAL ASSESSMENT

3. **Assessment of Risks**

The degree of risks in vulnerable areas needs to be assessed in order to provide a firm basis for development plan policies and proposals. The following procedure can be adopted:

a. have vulnerable areas been defined?	Comment:
b. has the potential for damaging events been defined?	Comment:
c. what is the existing standard of protection against damaging events?	Comment:
d. have the assets at risk been identified?	Comment:
e. have the potential consequences of damaging events been assessed?	Comment:
f. what is the likelihood of a damaging event?	Comment:
g. has the specific level of risk associated with an event of a particular size been assessed?	Comment:
Note: In many instances it may be inappropriate to evaluate risk in absolute terms because of the uncertainties involved. It may be more useful to assess the relative risk between a number of sites.	

Note: Flood risk information should be provided to local planning authorities by the NRA.

Coastal Defence Groups will probably have assessed future coastal cliff erosion rates as part of a Shoreline Management Plan.

4. **Review of Environmental Resources**

It is important to identify those environmental resources that are dependent on the continued operation of erosion, deposition and flooding processes. These may include (tick as appropriate):

EROSION	DEPOSITION	FLOODING
• geological features on coastal cliffs • mass movement sites • estuary habitats • natural succession in sand dunes • coastal cliff landscapes	• mudflats and saltmarshes • beaches and shingle ridges • sand dunes	• wetlands • mudflats and saltmarshes

Note: The relevant conservation agency should be consulted to identify how sensitive these features may be to the effects of development.

5   **Presentation of Information (Applied Earth Science Mapping)**

The accessible presentation of information is an important part of the general assessment process. Thematic maps can be prepared to support decision making. A wide range of maps may be relevant for specific audiences in different areas, as indicated below.

	MOUNTAINOUS AND UPLAND AREAS	UNDULATING LOWLANDS	ESTUARIES	LOWLANDS	CLIFFS	SAND DUNES
Element Maps						
Geology	•	•	•	•	•	•
Superficial Geology	•	•	•	•	•	•
Made Ground		•	•	•		
Geomorphology	•	•	•	•	•	•
Slope Steepness	•				•	
Soils	•	•	•	•		
Hydrogeology	•	•	•	•	•	•
Hydrology	•	•	•	•		•
Mineral Workings	•	•	•	•	•	•
Caves/Sink Holes	•	•			•	
Offshore Sediments			•	•	•	•
Derivative Maps						
Landslide Hazard	•				•	
Flood Potential	•	•	•	•		•
Erosion Potential	•	•			•	•
Geotechnical Conditions	•	•	•	•	•	•
Potential Mineral Resources	•	•	•	•	•	•
Sediment Budget			•	•	•	•
Sensitivity	•	•	•	•	•	•
Summary Maps						
Resources	•	•	•	•	•	•
Constraints	•	•	•	•	•	•

EROSION, DEPOSITION AND FLOODING

SITE REVIEW

Guidelines for Earth Science Considerations in the Preparation of Development Plans

These guidelines have been prepared to assist those involved in allocating land for specific purposes during the preparation of development plans. Local planning authorities should establish whether potential sites can be safely developed without the need for coastal defence improvements.

1.　　**Identification of Vulnerable Settings**

In considering the occurrence of vulnerable areas, the following should be consulted.

Flooding	Estuary Erosion and Deposition	Coastal Erosion
• NRA	• NRA	• Coast Protection Authority
• Internal Drainage Board	• Internal Drainage Board	• Coastal Defence Group
• Local Authority Coastal Engineers		

TERRAIN UNIT	Y/N	KEY PROCESSES AND VULNERABLE SETTINGS (tick)		IS THIS A SIGNIFICANT PLANNING CONSTRAINT?
Mountainous and Upland Areas		Upland Soil Erosion	• Upland peaty soils • Areas of recreation pressure • Recently afforested land	
		Flash Floods	• Land adjacent to streams • Areas downstream of dams and reservoirs	
		Sedimentation	• Upland lakes and reservoirs	
Undulating Lowlands		Water Erosion	• Areas of poor land management associated with silty and fine sandy soils	
		Wind Erosion	• Areas of poor land management associated with fine sandy soils	
		Flash Floods	• Land adjacent to tributary streams, especially around the margins of upland areas. • Areas downstream of dams and reservoirs	
		Unstable River Channels	• Alluvial rivers around the margins of upland areas • Alluvial rivers upstream and downstream of channelisation, river engineering works, bridges etc.	
		Mudfloods	• Dry valleys in areas of poor land management	
		Lowland Floods	• Low lying land adjacent to rivers and streams	
		Sedimentation	• Within alluvial river channels	
Estuaries		Flooding	• low lying land	
		Channel Erosion	• alluvial river channels	
		Sedimentation	• mudflats and saltmarshes • within channel	
Coastal Lowlands		Flooding	• low lying land, including land behind sand dunes shingle ridges etc.	
Coastal Cliffs		Landsliding and Cliff Recession	• soft rock cliffs	
Sand Dunes		Wind Erosion	• unstabilised bare sand areas	

2.   Flood and Coastal Defences

Where sites are at risk it is important to ensure that satisfactory coastal defence infrastructure exists and can be maintained over the life of the development or that new defences can be provided by the developer that do not affect other interests. The following procedure can be adopted:

a. what is the preferred land use or type of development?	Comment:
b. what is the standard of the existing flood or coastal defences?	Comment:
c. is this adequate for the preferred land use or type of development?	Comment:
d. can maintenance and improvements to defences be justified over the life of the development; or would defences need to be provided by the developer?	Comment:
e. would the development or the defences increase risks elsewhere?	Comment:
f. would the development or the defences affect conservation or mineral resources elsewhere?	Comment:
g. could potential problems (e and f) be overcome? If so, how?	Comment:
h. would the development and defences be consistent with Catchment Management Plan or Shoreline Management Plan policies?	Comment:

Note: These coastal defence issues should be discussed in consultation with:

- NRA
- Coast Protection Authority
- Local Authority Engineers
- Coastal Defence Group
- Conservation Agencies

3.   Nature and Scale of Flood and Coastal Defence Needs

Local plan development proposals should take due account of the physical constraints and the nature and scale of any precautionary measures that may be required. Prospective developers will thus be forewarned of the likely problems and can estimate the cost implications.

a. what target standard of protection would be required?	Comment:
b. what is the range of precautionary works that could meet the target standard of protection?	Comment:
c. in principle, would the range of precautionary works be acceptable on land use planning grounds?	Comment:
d. what are the possible cost implications of the precautionary works? Would these be economical for the developer?	Comment:

Note: These flood and coastal defence issues should be discussed in consultation with:

- NRA
- Coast Protection Authority
- Coastal Defence Group

EROSION, DEPOSITION AND FLOODING

Guidelines for Earth Science Considerations in the Determination of Planning Applications : Coastal Cliffs.

These guidelines have been prepared for the Department of the Environment by Rendel Geotechnics to those who determine planning applications.

1. **Checklist for Initial Considerations**

	YES	NO	COMMENT
● is the site in a risk area?			
● could the site affect a risk area?			
● have instability and erosion issues been brought to the attention of the developer?			
● was a stability report required?			

2. **Nature of the Stability Investigation**

INVESTIGATION TYPE	REQUIRED BY THE LOCAL PLANNING AUTHORITY?			UNDERTAKEN BY THE DEVELOPER		
	YES	NO	COMMENT	YES	NO	COMMENT
Desk Study						
Walkover Survey						
Surface Mapping						
Subsurface Investigation						
Laboratory Testing						
Monitoring						
Environmental Assessment						
Risk Assessment						

SUPPORT DATA	REQUIRED?	REVIEW OF REPORT INDICATES THAT		
		REPORT IS ACCEPTABLE?	ADDITIONAL DATA NEEDED?	COMMENT
● regional setting				
● surface geology				
● geomorphology				
● slope steepness				
● groundwater				
● hydrology				
● erosion rates				
● ground movement rates				
● history of landsliding				
● foreshore conditions				
● marine processes				
● littoral cell processes				
● stability analysis				
● sediment transport modelling				
● recession potential				
● landslide potential				
● environmental impact				
● risk assessment				

4.　　　**Review of Environmental Considerations**

RESOURCE	DO RESOURCES OCCUR IN AREA		REVIEW OF SUPPORTING INFORMATION INDICATES THAT					
	YES	NO	ARE POTENTIAL IMPACTS CONSIDERED?			IS ADDITIONAL DATA NEEDED		
			YES	NO	COMMENT	YES	NO	COMMENT
Groundwater								
Surface Water								
Agricultural Land								
Mineral Resources								
Landscape								
Nature								
Geological								
Archaeological								

AREA OF INTEREST		SITE REVIEW			IS THIS CONCLUSION DOCUMENTED IN ATTACHED REPORTS?		IS FURTHER SUPPORTING INFORMATION REQUIRED		
		YES	NO	COMMENT	YES	NO	YES	NO	COMMENT
Site	• is the land capable of supporting the loads to be imposed?								
	• could the development be threatened by instability within the site?								
	• could the development be threatened by instability on adjacent slopes?								
	• could the development be threatened by erosion during the lifetime of the building?								
Adjacent Land	• could the development initiate instability by unloading adjacent slopes?								
	• could the development initiate instability by loading adjacent slopes?								
	• could the development initiate instability by uncontrolled discharge of surface water?								
	• could the development initiate instability by uncontrolled discharge of foul sewerage?								
Broader Coastal Zone	• could the development increase erosion or flood risk elsewhere by reducing the supply of littoral sediment from cliffs?								
	• could the development increase erosion or flood risk elsewhere by disrupting the transport of littoral sediment along the foreshore?								

AREA OF INTEREST	QUESTIONS	YES	NO	PROPOSED MEASURES	ARE ADEQUATE DESIGN DETAILS PROVIDED?		WOULD PROPOSED MEASURES BE CONSISTENT WITH SHORELINE MANAGEMENT STRATEGIES		
					YES	NO	YES	NO	COMMENT
Site	• can stability conditions be improved?								
Adjacent Land	• can the effects of the development on adjacent slopes be mitigated?								
Broader Coastal Zone	• can the effect on natural movement of littoral sediment be minimised?								

7.   **Sources of Background Information**

1.	British Geological Survey geological maps.
2.	National Landslide Databank and Maps; held by Rendel Geotechnics for the Department of the Environment.
3.	Review of Landsliding in Great Britain: GEOMORPHOLOGICAL SERVICES LTD 1986-87 Review of research into landsliding in Great Britain. Rpts to Dept Environment Openfile reports. Series A Regional review of landsliding Vol I South East England and East Anglia Vol II South West England Vol III The Midlands Vol IV Wales Vol V Northern England Vol VI Scotland. Series B Causes and mechanisms of landsliding Vol I International review Vol II Britain. Series C (by RENDEL PALMER AND TRITTON) Landslide investigation techniques and remedial measures - Research and Practice 2 Vols. Series D Landslides and policy Vol I Landslide hazard assessment Vol II Landslide risk in Britain Vol III Legislative and administrative provisions. Practice in England and Wales and a review of overseas practice. Series E National summary and recommendations.
4.	Jones DKC and Lee EM 1994 Landsliding in Great Britain. HMSO.

Printed in the United Kingdom for HMSO
Dd301776 12/95 C8 G559 10170